DEPARTMENT OF HEALTH

Committee on the Medica

of Air Pollutants

Cover illustration - (CH 43767) Waterloo
Bridge, 1900 by Monet, Claude (1840-
1926) Christies Images/Bridgeman Art
Library, London

Monet painted a series of studies of
London in smog in the early years of this
century. The Department of Health is
indebted to Dr David Bates (Emeritus
Professor of Medicine, Vancouver) for the
idea of using one of Monet's paintings to
illustrate this Handbook.

DEPARTMENT OF HEALTH

Committee on the Medical Effects

of Air Pollutants

HANDBOOK ON AIR POLLUTION
AND HEALTH

LONDON: THE STATIONERY OFFICE

Printed in the United Kingdom for The Stationery Office.
Dd 304144 c13 8/97

Preface

Air quality is of interest to many people. Though air quality in the UK has improved very significantly since the smogs of the 1950s levels are still such as to produce effects on health. This book provides an introductory account of the effects of air pollutants on health. The information provided is based on the detailed reports prepared during the past seven years by the Department of Health's Advisory Group on the Medical Aspects of Air Pollution Episodes and the Committee on the Medical Effects of Air Pollutants. It is hoped that this book will make available, more widely, the work of these groups and will provide a ready source of advice and reference on the effects of air pollution.

Professor S T Holgate
Chairman
Committee on the Medical Effects of Air Pollutants

August 1997

All references to the Government in this Handbook relate to the Administration prior to the General Election of May 1997. The Department of the Environment is now part of the Department of the Environment, Transport and the Regions.

Contents

Page

Acknowledgements

The Department wishes to acknowledge those authors and publishers who have given permission for material to be reproduced in this Handbook.

The Department also wishes to acknowledge the expert assistance provided by Wordcraft (South Bank Technopark) in designing the layout of this Handbook.

SECTION 1

OVERVIEW OF AIR POLLUTION AND HEALTH

- Introduction
- Risk to health posed by air pollutants
- Effects of exposure to combinations of air pollutants

OVERVIEW OF AIR POLLUTION AND HEALTH

INTRODUCTION

1.1   Since the dawn of history, mankind has been burning biological and fossil fuel to produce heat. Heating for homes and businesses, the generation of electrical power and many other uses has inevitably led to the release of the products of combustion and pollution of the air. Recognition that air pollution could be damaging to health and should be reduced occurred in England in the 13th century, though many years passed before progress in pollution control was made. During the period of the Industrial Revolution in European countries and elsewhere, pollution was seen as an inevitable accompaniment of industrial development and some early efforts were made to minimise its possible effects on health.

The first English account of air pollution 1661

FUMIFUGIUM:

OR,

The Inconvenience of the AER,

AND

SMOAKE of LONDON

DISSIPATED

TOGETHER

With some REMEDIES humbly proposed

By John Evelyn Esq;

To His Sacred MAJESTIE,

AND

To the PARLIAMENT now Assembled.

Published by His Majesty's Command.

Lucret. l. 5.

Carbonumque gravis vis, atque odor insinuatur.
Quam facile in Cerebrum?——

1.2   During the Twentieth Century, the perception that air pollution damages health has grown and during the last fifty years many countries have introduced laws to reduce pollution. However, sources of air pollution have changed. Whilst the use of coal for domestic heating was once the dominant source of pollution in cities, now the major concern in most Western European cities is exhaust from motor vehicles.

1.3   The study of the effects of air pollution on health began with the study of acute and severe air pollution episodes. Three episodes in particular focused attention on air pollution: an episode in the Meuse Valley, Belgium in 1930, an episode in Donora in the USA in 1948 and an episode in London, the UK, in 1952. These episodes were studied in detail and an estimate of the number of deaths attributable to the increased levels of air pollution (the excess deaths) made. The following table shows the number of excess deaths associated with some of the major air pollution episodes this century.

Table 1.1.
Major Air Pollution Episodes Occurring in the Twentieth Century

Date	Place	Excess deaths
December 1930	Meuse Valley, Belgium	63
October 1948	Donora, USA	20
December 1952	London UK	4700
November 1953	New York, USA	250
January 1956	London UK	480
December 1957	London UK	300-800
Nov-Dec 1962	New York USA	46
December 1962	London UK	340-700
December 1962	Osaka, Japan	60
Jan-Feb 1963	New York, USA	200-405
November 1966	New York USA	168
December 1991	London UK	100-180

(Based on Elsom D: Atmospheric Pollution, published by Blackwell, Oxford UK, 1987)

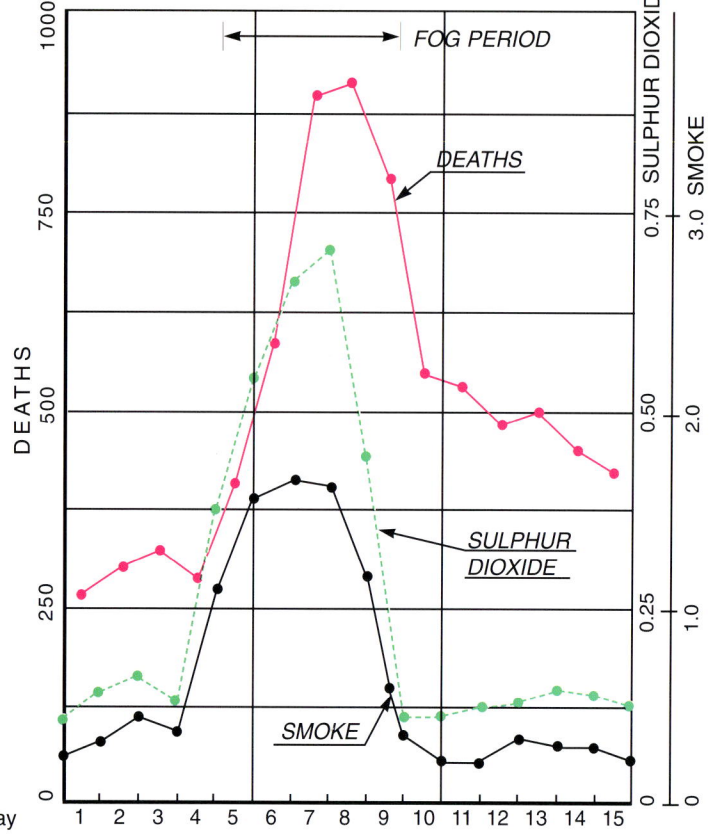

Fig 1.1
Deaths and Air Pollution during
December 1952 in Greater
London.

Sulphur Dioxide: ppm
Smoke: mg/m³
Deaths: Number per day

It can be seen that all the episodes listed, with the exception of the Donora incident, occurred in the middle of winter. Again, with the exception of the London 1991 episode, all the others were characterised by high concentrations of smoke and sulphur dioxide. The 1991 London episode was characterised by high levels of nitrogen dioxide and particles. In earlier analyses of the effects of air pollution episodes on health emphasis was placed on the excess deaths occurring as a result of episodes. More recently, attention has focused on other effects including increased admissions to hospital, use of medicines, and reduction in personal activity.

Sources of air pollutants

1.4   Burning of fossil fuels produces carbon dioxide, water and a range of other compounds. Coal, for example, contains sulphur and sulphur dioxide is inevitably produced when it is burned. Benzene, polycyclic aromatic compounds, methane, formaldehyde and others are produced by burning petrol in an internal combustion engine. At high temperatures, atmospheric nitrogen is oxidised to nitric oxide (NO) and further to nitrogen dioxide (NO_2) and these are also released to the air.

1.5   The contribution of different sources to the total output of air pollutants varies from country to country and from area to area. The advances made in the UK in the years following the Clean Air Act of 1956 depended on the establishment of local smoke-free zones and the conversion of old fashioned and inefficient domestic fires either to gas or electricity or to a type which could burn smokeless fuel.

The success of these national policies, implemented and enforced at a Local Authority level, can be seen in every major city in the UK, although there are still areas where coal is burned. Similar approaches adopted in many other European countries have led to a general improvement in the urban environment and black buildings, smoking chimneys and coal smoke smog are now largely things of the past. However, this old fashioned smog has been replaced by a newer and different form of air pollution due largely to the increased dependence on motor vehicles as the major means of transport. Thus one air pollution problem has been replaced by another, perhaps more intractable one. In some countries, both problems exist and air pollution then poses perhaps its most serious threat to health. This is true of some large cities in developing countries where air pollution is now very severe.

Domestic use of coal as a source of pollution

1.6   Coal is a significant source of sulphur dioxide emissions and other pollutants including particles, nitrogen dioxide, carbon monoxide, organic compounds such as polycyclic aromatic hydrocarbons and various inorganic compounds. It has been one of the most important sources of air pollution in the past. But while this has improved in recent years, solid fuels are still widely used in certain parts of the UK such as Belfast which, currently, has no piped natural gas.

1.7   The amounts of pollutants produced depends on the composition of the coal. For example, anthracite contains as much as 95% carbon and produces smaller amounts of pollutants than other coals though it is a major source of the greenhouse gas carbon dioxide. Special devices with carefully controlled air supplies are needed to burn anthracite (it does not burn easily on open fires) but combustion is efficient and far less smoke is produced.

Power generation as a source of pollution

1.8   The policy to move fossil fuel-burning power stations from urban to rural areas and to distribute the pollution produced more evenly via very tall chimneys has led to an improvement in urban air quality. Better dispersion of pollutants emitted by tall chimneys leads to better dilution in the air and thus lower local concentrations. However, non-nuclear power stations remain an important source of some pollutants. For example in the UK over 90% of sulphur dioxide released to the air is produced by industrial sources with about 70% being produced by the power generating industry. Coal and oil contain sulphur (in the UK, coal about 1.7% and heavy fuel oil up to 3% by weight). Much can be done to reduce the output of sulphur dioxide from power stations by fitting desulphurisation equipment and the countries of the European Union are committed to major reductions in sulphur dioxide output. Power stations also produce particulate materials (fly ash) (which can again be reduced by electrostatic precipitators and filters) and oxides of nitrogen (NO_x). Current estimates show that on a national scale, whilst road transport is the main source of emissions of NO_x, accounting for 49% of the total, power stations also make an important contribution, producing approximately 24% of the total UK emissions of NO_2.

Motor vehicles as a source of air pollution

1.9   Traffic growth in many countries has been, and is, rapid and until recently little attention has been paid to the air pollution generated. The motor car has

brought notable advantages in terms of personal independence and mobility but against these should be set the effects on the environment. It seems unlikely that a significant reduction in the popularity of the motor car will occur for some time but efforts to plan car use, to further improve emission control and to inform motorists of the cost to the environment resulting from the use of motor cars are needed to control increasing levels of pollution. In many countries such steps have been taken and a progressive tightening of emission standards has been a policy in European countries, including the UK, for some years. The removal of lead from petrol made possible the use of effective and economically viable catalytic converters. These reduce emissions of carbon monoxide, nitrogen dioxide and volatile organic compounds and have made the properly maintained and sensibly driven modern motor car a remarkably "clean" vehicle. Alternative modes of transport are being developed and are being adopted increasingly.

2nd QUARG report (see ref list)

1.10   The success of catalytic converters has put pressure on diesel vehicles and the perception that diesel powered cars are "cleaner" and more "eco-friendly" than petrol powered cars has changed. Compared with catalyst-equipped petrol cars, diesel cars emit less carbon monoxide and hydrocarbons but more particles and nitrogen dioxide. Diesel powered vehicles are an important source of particulate material and concern about the effects of fine particles upon health has put pressure on the manufacturers of these vehicles. Of course, no vehicle burning fossil fuel can be completely non-polluting. At best, carbon dioxide, water and nitrogen are produced. Carbon dioxide is an important green-house gas and a reduction in consumption of fossil fuels is needed if global warming is to be reduced.

Diesel Vehicle Emissions and Urban Air Quality

1.11   The air pollutants produced from motor vehicles can be divided into primary and secondary pollutants.

- Primary pollutants are those released directly into the air from the pollutant source and, for petrol powered vehicles, include carbon monoxide, nitric oxide, benzene and particulate material. Much of the lead emitted by vehicles burning leaded petrol emerges as particles. Diesel engines burn fuel in an excess of air and so produce little carbon monoxide.

- Secondary pollutants are those formed by chemical changes to the primary pollutants. For example, nitrogen dioxide is produced from the nitric oxide emitted by motor vehicles and thus is a secondary pollutant. Photochemical breakdown of nitrogen dioxide leads to the formation of the secondary pollutant ozone.

Fig 1.2

Production of ozone in the troposphere

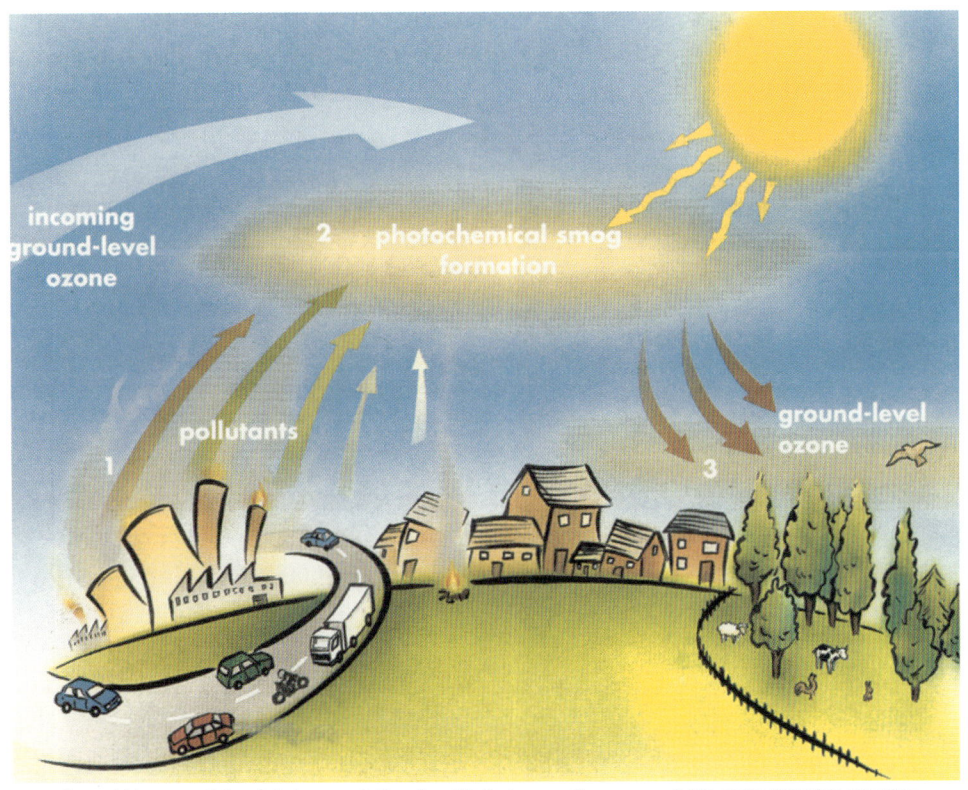

1. Motor vehicles, power stations, industry and domestic sources produce pollutants. These rise into the lower atmosphere.

2. The action of the Sun's rays on these pollutants produces photochemical smog, including ground-level ozone.

3. This reaction takes place over quite a time and distance, so much ground-level ozone eventually reaches rural areas.

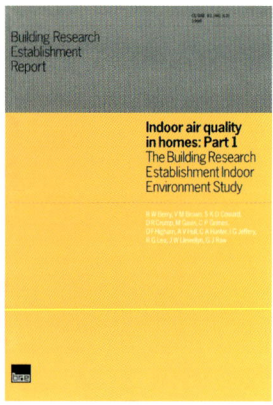

Indoor Air Quality in the UK
BRE Report 1996
(see ref list)

Risks to Health posed by Air Pollutants

1.12   One of the major concerns with air pollution is the risk to health. It is possible that this particular area has been over-stressed in relation to other challenges in the field of public health. Cigarette smoking has a much more significant effect on the prevalence of many respiratory diseases than air pollution. Extensive research in many countries has helped determine the risk from acute air pollution episodes but there are still considerable areas of uncertainty, particularly with respect to long term health effects. Distinguishing the effects of air pollution from those of other factors such as cigarette smoking is not easy.

1.13   Much attention is directed towards outdoor air pollution, although most people in the UK spend more than 80 percent of the time indoors. For some pollutants, indoor sources can make a far greater contribution to total exposure than outdoor sources, partly because of the amount of time spent indoors but also because of emissions from indoor sources such as nitrogen dioxide from gas cookers.

1.14   As efforts have been made to improve energy efficiency through better insulation and reduced ventilation of domestic dwellings, so the opportunity for accumulation of pollutants within homes has increased. The introduction of new building products and furnishings within the home can also have an impact on indoor air quality. It is important, therefore, that any assessment of the effects of air pollution on health takes into account both exposures outside and within the home. For this reason, further research on the effects of indoor air on health is being encouraged. Some of the major indoor air pollutants are covered in Section 3, but the main emphasis of this handbook is on outdoor air pollution. There is also no attempt to consider the effects of exposure to radio-isotopes such as radon, which is an important indoor air pollutant in some areas in the UK.

Main health effects of air pollution

1.15   The following sections summarise briefly the main health outcomes which are understood to be affected by air pollution, and the evidence associated with these. More detailed notes on the effects of the individual air pollutants are in the sections dealing with individual pollutants. Air pollution can cause acute effects on health close to the time of an air pollution episode and, possibly, more long term effects due to exposure to air pollution over a long period of time. The latter is more difficult to study and more uncertain. Air pollution is most usually associated with respiratory diseases, though effects may be most marked in those suffering from chronic combined heart and lung disease. The factors affecting acute respiratory illness are different from those affecting cancer.

Chronic bronchitis and emphysema

1.16   The major cause of these diseases is cigarette smoking but there is some evidence that atmospheric pollution may also contribute. Since the 1950s there has been evidence showing that these diseases occur more commonly in urban areas. There is also evidence that chronic bronchitis and emphysema are aggravated by smoke, sulphur dioxide, and other pollutants, and that patients suffering from these diseases are likely to be less well during air pollution episodes. Recent studies have shown that the risk of death amongst such patients is increased as concentrations of air pollutants rise.

Cardio-respiratory death

1.17   In addition to patients suffering from heart disease alone, patients with chronic bronchitis and emphysema often have associated heart problems. Cross-sectional epidemiological studies from North America of health effects associated with particle concentrations have shown an increase in risk of cardio-respiratory death with increasing particle concentration. However, it is difficult to be sure how far these results reflect other confounding factors such as socio-economic differences, between different areas. The incidence of heart attacks (myocardial infarctions) has been shown in a recent study to increase on days when concentrations of particles are raised. Environmental exposure to carbon monoxide may also be related to episodes of heart failure.

Susceptibility to respiratory infection

1.18   In the indoor environment, there is concern that nitrogen dioxide from gas cookers may adversely affect lung function and increase susceptibility to respiratory infection in children.

1.19   There is now good evidence to show that infants and children exposed to Environmental Tobacco Smoke (ETS) experience a higher incidence of respiratory symptoms, Sudden Infant Death Syndrome (SIDS) and asthma. There are also concerns about chronic effects as well as possible links with cardiovascular disease and effects on pregnancy.

Lung and other cancers

1.20   The major cause of lung cancer is cigarette smoking which accounts for 90% of all lung cancers. Because of the overwhelming effect of cigarette smoking it is difficult to assess any contributory role of air pollution which is clearly small in relation to smoking. Environmental (passive) cigarette smoke has been estimated to cause several hundred lung cancer deaths per year. Radon, an indoor pollutant in some areas of the country, has known carcinogenic effects and may also contribute to deaths from lung cancer.

1.21   A considerable number of carcino-genic substances occur as outdoor, and indoor, air pollutants. Of these, benzene, 1,3-butadiene and the polycyclic aromatic hydrocarbon compounds (PAHs) are perhaps the best known. Some are genotoxic carcinogens and could therefore cause an increase of cancer even at low

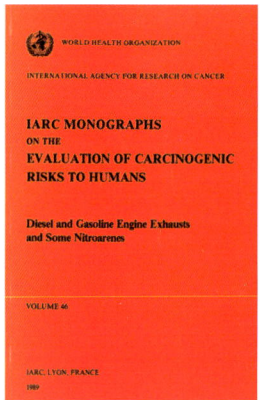

Report on diesel emissions - by IARC 1989

levels of exposure. The term "cancer" is used here in its widest sense to embrace all malignant tumours and related disorders such as leukaemia. Attempts to quantify the risk posed by exposure to ambient levels of carcinogens have been made though the assumptions underlying these calculations have often been challenged as described below.

1.22 Indoor air quality can also be affected by the land on which the building stands and materials used for construction and furnishing. The NRPB estimate that radon inside buildings causes about 2000 lung cancer deaths per year in the UK. This is a particular problem in parts of the country where the underlying rock is rich in uranium.

1.23 In 1989, the International Agency for Research on Cancer (IARC) classified diesel emissions as probably carcinogenic to humans. This conclusion was based on sufficient evidence of the carcinogenicity of whole diesel engine exhaust and extracts of diesel exhaust particles in animals and limited evidence of the carcinogenicity of diesel engine exhaust in humans. Sufficient evidence of carcinogenicity is defined by IARC as follows:

> The Working Group considers that a causal relationship has been established between the agent or mixture and an increased incidence of malignant neoplasms or of an appropriate combination of benign and malignant neoplasms in (a) two or more species of animals or (b) in two or more independent studies in one species carried out at different times or in different laboratories or under different protocols.

In 1990, the Department of Health's Committee on Carcinogenicity of Chemicals in Food, Consumer Products and the Environment (COC) concluded that lifetime exposure of rats to very high inhaled concentrations of whole diesel exhaust led to an increased incidence of benign and malignant lung tumours. Epidemiological data indicated that sustained long-term exposure to diesel exhaust at high occupational levels was associated with an increased incidence of lung cancer. The possibility of a small increased risk of lung cancer due to general environmental exposure to diesel exhaust could not be excluded on the evidence available at that time. A subsequent consideration of animal studies by COC in 1996 concluded that the carcinogenic effects of diesel exhaust in animal studies appeared to be specific to the rat and resulted from progressive impairment of the clearance mechanisms that normally prevent accumulation of particles within the lung. The COC concluded that these gross overload effects were not relevant to the assessment of risk to humans exposed to diesel exhaust and did not allow conclusions to be made regarding the mechanisms of tumour induction by diesel particles in humans.

1.24 IARC also considered petrol emissions in 1989. Petrol emissions were classified as possibly carcinogenic to humans. This conclusion was based on sufficient evidence of the carcinogenicity of petrol-engine exhaust condensates/extracts in animals although the evidence of carcinogenicity in animals and humans of whole petrol-engine exhaust was inadequate.

1.25 An epidemiological study from the US has shown a statistical relationship between long term levels of particles and mortality, with lung cancer being one of the causes of increased mortality. A further study of 151 cities has confirmed this finding.

Other health effects

1.26 About 60 deaths occur each year in homes due to accidental carbon monoxide exposure from faulty gas appliances. Also, unknown, but possibly large numbers of people experience sub-lethal exposures.

Air pollution, asthma and allergies

1.27 In asthma the airways are unusually sensitive to a wide range of stimuli, including inhaled irritants and allergens. This results in obstruction to airflow which is episodic - at least in individuals with early or mild asthma - and which causes symptoms of tightness and wheeziness in the chest. There has been an increase of about 50% in the prevalence of childhood asthma in the UK over the last 30 years, associated with an increase in other allergic diseases, such as hay fever. There has been at least a ten-fold increase in hospital admissions for asthma among children, although this, in part, reflects changes in medical practice. Significant increases in the prevalence of asthma have also occurred in other countries. Over the period during which asthma has been increasing, emissions of coal smoke and sulphur dioxide have fallen markedly in the UK and in many West European countries, while those of oxides of nitrogen and volatile organic compounds from motor vehicles have increased. During this time emissions of particles from coal fires have fallen, whilst those from diesel vehicles have increased.

This special issue of Clinical & Experimental Allergy vol 25, Supp 3, 1995 discussed the role of air pollution in asthma and was funded by DH.

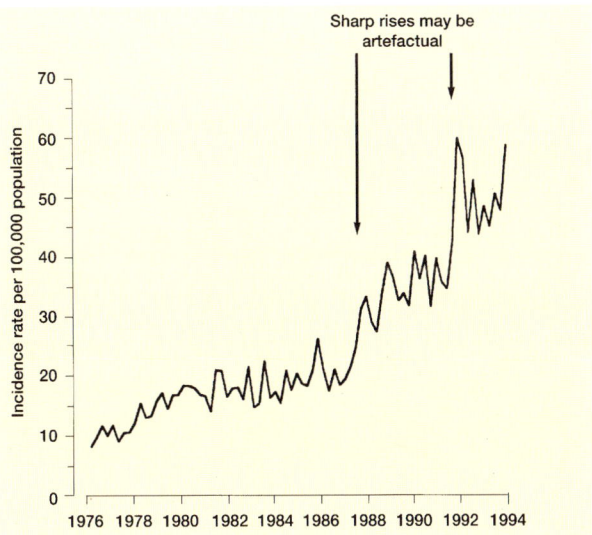

Fig 1.3
Trends in Asthma Reported by GPs - Mean weekly incidence in 12 week periods - All persons all ages England and Wales 1976 - 1993

1.28 It has been suggested that environmental factors such as air pollution could initiate asthma in previously healthy individuals or provoke or aggravate asthma symptoms in those who are already asthmatic. While there is laboratory evidence that air pollution could potentially have a role in the initiation of asthma, there is no firm epidemiological or other evidence that this has occurred in the UK or elsewhere. While there is some epidemiological evidence that air pollution may provoke acute asthma attacks or aggravate existing chronic asthma, the effect, if any, is generally small and the effect of air pollution appears to be relatively unimportant when compared with several other factors (eg, infections and allergens) known to provoke asthma.

Fig 1.4
Prevalance of asthma in British children in urban, mixed and rural areas

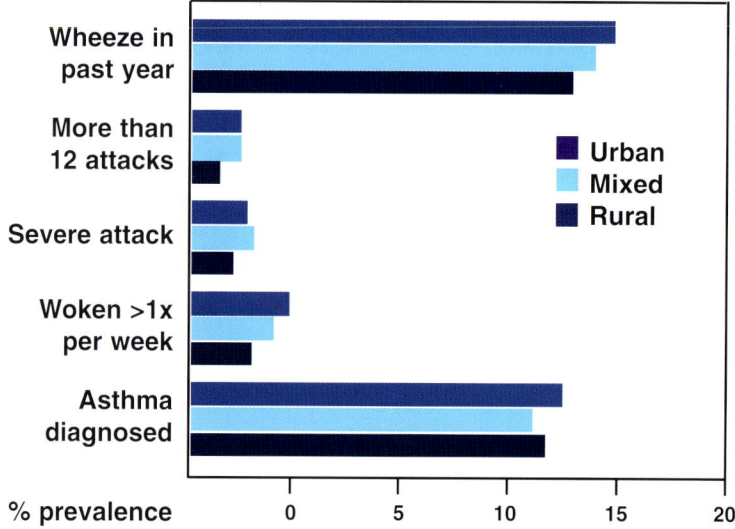

Data taken from D.P. Strachan et al., (1994) Arch.Dis.Child.,70; 174-178

1.29   Data on trends and geographical variations in exposure to ozone, nitrogen dioxide and particles from vehicles are limited. The occurrence of ozone episodes in summer, in the UK, has probably increased over this century, but in the 15-20 years since measurements began, there is no clear trend in annual average concentrations. Annual average nitrogen dioxide concentrations have not increased in large urban centres in the UK, although there is some indication of a small increase in other urban areas of the UK.

1.30   There is some laboratory evidence that exposure to the common gaseous pollutants can enhance the response of asthmatic patients to allergens though the effect does not seem to be large. There is no direct evidence for such an interaction as a result of exposure to outdoor air pollution in the UK.

1.31   There is no consistent relationship between trends in the prevalence of asthma and trends in emissions or ambient concentrations of air pollutants. A number of equally, if not more, plausible explanations for the trends in asthma have been hypothesised, a reduction in early childhood infections receiving particular attention. The epidemiological evidence concerning the short term effects of air pollution on asthma indicates that:

(i)   Day-to-day variations in air pollution are likely to have a small effect on the lung function of asthmatic adults and children. In general these changes are unlikely to cause symptoms. However, patients with severe asthma may be more affected because of their lower reserve of lung function. The main effects are observed in the elderly with chronic obstructive lung disease (which includes asthma).

(ii) Seasonal patterns of asthma attacks bear little or no relationship to seasonal patterns of air pollution.

(iii) Based on studies from overseas, it is likely that the short-term fluctuations in levels of air pollution currently encountered in the UK are responsible for small changes in the numbers of hospital admissions and accident and emergency attendances for asthma. Limited experience from the UK during well defined air pollution episodes indicates that admissions may be increased by a small amount along with similar increases in admissions for other respiratory diseases.

1.32 The epidemiological evidence concerning the geographical distribution of asthma indicates that:

(i) There is little or no association between the regional distribution of asthma and that of air pollution.

(ii) Prevalence studies comparing high with low pollution areas have not found consistent associations between outdoor air pollution and asthma prevalence.

(iii) There is no convincing evidence that asthma is more common in urban areas than in rural areas of the UK. Limited evidence from the UK and other countries suggests a modest relationship between asthma prevalence and local traffic density. The extent to which this is due to air pollution has yet to be determined.

1.33 In summary, while there is evidence that air pollution can affect the frequency and severity of attacks in those who suffer from asthma, there is very little support for the suggestion that asthma is caused by air pollution, in particular by outdoor air pollution.

Magnitude of the effects of air pollutants upon health

1.34 The evidence relating day-to-day variations in concentrations of air pollutants to effects on health is stronger than that dealing with the effects of chronic exposure to long-term average concentrations.

1.35 The results of studies of the effects on health of day-to-day variations in concentrations of pollutants have been used to estimate overall, eg annual, effects of certain air pollutants. These estimates have led to assertions that the deaths of thousands of people are affected, each year, by exposure to particles. These calculations involve the assumption that there is no safe level of exposure to particles. This remains unproven.

1.36 One of the major problems in assessing the risk posed by air pollution is that we do not know the contribution which exposure to pollutants may make to deaths from, say, cardiovascular disease. In many countries cardiovascular disease is a leading cause of death and even a small effect of air pollution could mean a significant and important effect on public health.

1.37 Another important concept is that exposure to pollutants may precipitate deaths which would have, in any case, occurred very soon. This does not affect the numbers of deaths brought about by air pollution, but does affect their public health importance. If we knew that the maximal effect of exposure to ambient levels of particles was to bring death forward by only a few days then the effect might be considered less important. Unfortunately we have little knowledge of the extent to which deaths are hastened by exposure to pollution and so the question of the importance of the effect remains unresolved.

1.38 Can we then say anything valid and useful about the risk to health posed by air pollution? On an individual level the risk is very much smaller than that posed by active cigarette smoking or accidents. It is also true that healthy individuals are unlikely to be affected by exposure to concentrations of pollutants as occur outdoors on most days of the year. The old and the very young, especially those with pre-existing chronic cardio-respiratory diseases, are the groups who are vulnerable to the effects of air pollution.

Effects of exposure to combinations of air pollutants

1.39 Air pollution episodes are characterised by raised concentrations of more than one pollutant. In the UK, the air pollution episodes tend to have characteristics varying with season. In winter and in urban areas, the mixture is dominated by oxides of nitrogen, in summer, generally in more rural areas, ozone dominates the mixture. High levels of particles can occur in any of these episodes. The possibility of additive or even synergistic effects is obvious and the DH Advisory Group on the Medical Aspects of Air Pollution Episodes (MAAPE) was asked to examine this in their fourth and last report.

1.40 This was published in November 1995. MAAPE concluded that there was no clear evidence of synergism, although effects of pollutants in mixtures could be additive. MAAPE further concluded:

> It is probable that all three main types of pollution mixtures encoun-
> tered during air pollution episodes in the United Kingdom could cause
> small mean reductions in lung function in normal individuals. The
> evidence for this is probably strongest for ozone-related pollution,

where reduction in lung function and development of symptoms is more likely in those who exercise. There is no evidence that any of the three types of episode commonly seen in the United Kingdom cause symptoms or adverse health effects in people who are otherwise well.

It is likely that episodes of air pollution occurring in the United Kingdom produce adverse health effects in some persons with chronic respiratory disease. There is also evidence that some individuals with asthma might experience some deterioration in their condition, although the majority have shown little effect in most studies.

The elderly, especially those with chronic heart or lung disorders, are identified as being at increased risk of deterioration [of their clinical state] during air pollution episodes. The degree of increased risk incurred by these groups is difficult to quantify.

It might be expected that patients with asthma would be especially vulnerable to air pollution. At levels encountered in the United Kingdom this has proved difficult to demonstrate convincingly, except possibly in those with severe asthma. It is possible that asthmatic patients might compensate for any effects of air pollutants by increasing their medication.

COMMITTEES ADVISING GOVERNMENT ON AIR POLLUTION

- **MAAPE**
- **COMEAP**
- **QUARG**
- **EPAQS**

Committees advising Government on Air Pollution

DH Advisory Group on the Medical Aspects of Air Pollution Episodes (MAAPE)

2.1   In 1990 the Chief Medical Officer (CMO) established the Advisory Group on the Medical Aspects of Air Pollution Episodes (MAAPE). This group was specifically asked to examine the likely effects of exposure to such episodes of elevated levels of air pollutants as occur in the UK and to advise CMO regarding the need to provide advice to the public of these effects and if necessary of what measures might be taken to ameliorate them. The group has now been subsumed into COMEAP.

Reports: **Ozone (August 1991)**
          Sulphur dioxide, acid aerosols and particulates (October 1992)
          Oxides of nitrogen (December 1993)
          Health effects of exposures to mixtures of
          air pollutants (November 1995)

DH Commitee on the Medical Effects of Air Pollutants (COMEAP)

2.2   In 1992, COMEAP was created to provide a broader assessment of the public health effects of air pollutants than that provided by MAAPE. Also there was a need for regular input to the Expert Panel on Air Quality Standards.

Reports: **Asthma and Outdoor Air Pollution (October 1995)**
          Non-Biological Particles and Health (November 1995)

The terms of reference of COMEAP are:

At the request of the Department of Health:

(a) to assess and advise Government on the effects upon health of air pollutants, both outdoor and indoor air, and to assess the adequacy of the available data and the need for further research, and

(b) to coordinate, with other bodies concerned with the assessment of the effects of exposure to air pollutants and their associated risks to health, and to advise on new scientific discoveries relevant to the effects of air pollutants upon health.

Quality of Urban Air Review Group (QUARG)

2.3   QUARG was set up by the Department of the Environment in 1992. The group has produced three reports:

Urban Air Quality in the United Kingdom   **1993**
Diesel Vehicle Emissions and Urban Air Quality   **1993**
Airborne Particulate Matter in the United Kingdom **1996**

Each report has dealt with its subject in depth and has provided important material for the development of the United Kingdom Air Quality Strategy.

DoE Expert Panel on Air Quality Standards (EPAQS)

2.4   Following the White Paper 'This Common Inheritance', published in 1990, an Expert Panel was established to advise on air quality standards in the United Kingdom.   The panel began work in 1992.   One of the major factors being taken into account by the Expert Panel is the health effects of air pollutants with a health input from the Committee on the Medical Effects of Air Pollutants (COMEAP), and on more specialist areas from the appropriate DH expert Committees, such as the Committee on Carcinogenicity.

Reports: **Benzene (February 1994)**
Ozone (May 1994)
1,3-Butadiene (December 1994)
Carbon monoxide (December 1994)
Sulphur dioxide (September 1995)
Particles (November 1995)
Nitrogen dioxide (December 1996)

In March 1997, the National Air Quality Strategy was published, in which the Government adopted the recommendations of EPAQS as national air quality standards. Where the Expert Panel had not made a recommendation, the relevant information from the World Health Organisation was used, where available. The standards were used as benchmarks or reference points for setting national air quality objectives.

ADVICE FROM COMEAP

Health advice during air pollution episodes

2.5   COMEAP has endorsed two statements of advice for the Department of Health. These are intended for use during "winter" and "summer" air pollution episodes, the typical mix of pollutants being different in the two seasons:

WINTER AIR POLLUTION EPISODE

The Department of the Environment has indicated that an air pollution episode is to be expected during [the next 24 hours]. During this period levels of vehicle-related and other pollutants are expected to rise.

Most people will experience no ill effects; but those suffering from any heart or lung disease (including asthma), particularly if elderly, should be aware that their symptoms may worsen and should consult their doctors if this occurs.

People with asthma and others who are affected may need to consider modifying their treatment, consulting their doctors as necessary. Because winter   pollution   episodes   usually   occur   in   cold,   still   weather conditions, elderly people are advised to spend less time out of doors and keep warm. People who have in the past noticed that their breathing is affected by traffic fumes should avoid busy streets during the episode as far as is possible.

SUMMER AIR POLLUTION EPISODE

This advice is issued as part of a DoE press release when poor air quality is predicted during the summer months.

During episodes of air pollution experienced during the summer in the United Kingdom, levels of ozone, nitrogen dioxide and particles may be raised.

Most people will experience no ill effects; but those suffering from heart or lung disease (including asthma) should be aware that their symptoms may worsen. Those who are affected may need to consider modifying their treatment, consulting their doctors as necessary.

People who have noticed in the past that their breathing is affected on hot, sunny days should avoid strenuous outdoor activity, particularly in the afternoon. Children with asthma should be able to take part in games in the usual way, although they may need to increase their use of reliever medicines before participating. There is no need for them to stay away from school.

REVISION OF THE AIR QUALITY INFORMATION SERVICE

The Department of the Environment is currently considering revising the UK Air Quality Information Service. As a part of this process the Committee on the Medical Effects of Air Pollutants has examined the banding system for air quality and has proposed a number of revisions. These take into account the latest information of the effects of air pollutants on health and the recommendations of the DOE Expert Panel on Air Quality Standards (EPAQS). The Committee on the Medical Effects of Air Pollutants has prepared a statement on the revisions proposed. This is reprinted in Appendix 4.

INDIVIDUAL POLLUTANTS

- **Sulphur Dioxide**
- **Nitrogen Dioxide**
- **Ozone**
- **Particulate Matter**
- **Volatile Organics**
- **Benzene**
- **1,3-Butadiene**
- **Polycyclic Aromatic Hydrocarbons**
- **Carbon Monoxide**
- **Lead**
- **Cadmium**
- **Environmental Tobacco Smoke**
- **Formaldehyde**

INDIVIDUAL POLLUTANTS

Hazard, risks and standard setting

3.1   The terms hazard and risk are used by toxicologists in a specific way. "Hazard" denotes or describes the inherent toxicity of a compound. "Risk" takes account of the assessment of hazard but also includes an estimate of the likelihood and extent of exposure in reaching an estimate of the probability of effects occurring. Thus hydrocyanic acid is undoubtedly a compound which presents a considerable hazard though the risk of exposure to this compound in ambient air is very low.

3.2   An assessment of the hazard associated with exposure to a toxic compound will involve an account of the effects of the compound including the symptoms and clinical signs produced. This may be derived from studies in animals, human volunteers or as a result of observing the effects of accidental exposure to the compound. An assessment of risk, on the other hand, will comprise an estimate of the likelihood of adverse effects occurring.

3.3   The results of a risk assessment exercise may be expressed in numerical terms or in words. For example, the increased risk of developing leukaemia as a result of lifelong exposure to a level of 1 μg/m^3 benzene has been calculated by some authorities as 4×10^{-6} (4×10^{-6} means "4 per million" or 1 in 250,000). This type of mathematical estimation of risk is often referred to as "Quantitative Risk Assessment" (QRA) and has been applied mainly to compounds known or believed to produce cancer in man. In particular it has been applied to those compounds believed to damage the genetic material of the cell: genotoxic carcinogens.

3.4   The process usually involves extrapolation from data collected either in studies of experimental animals or of workers exposed to the chemical in question. Inherent in the process is the assumption that exposure to even very low concentrations will produce some increase in risk and that this can be predicted from some assumed dose response relationship. If the exercise is based on animal studies then the assumption that man will respond in a way similar to that of the animals is also made. That either of these key assumptions may be false has been pointed out by many commentators. The use of different models for extrapolation may also lead to differing estimates of risk. The application of such models to data collected from human epidemiology studies is generally regarded as more satisfactory than their application to data collected from animal studies.

3.5   Perhaps the greatest difficulty associated with QRA is that these assumptions cannot be tested. Clearly, experimental exposure of people to cancer-producing chemicals is impossible. Confirmation of the validity of the assumed exposure-response relationship at low levels of exposure in animals is also in practice impractical as it would involve the use of impossibly large numbers of experimental animals.

3.6   Another defect of QRA is that the results are often misused and taken to imply a level of accuracy or certainty which the original authors of the work certainly would not claim.

3.7   These problems have led regulatory toxicologists in the UK to be cautious regarding QRA. In other countries, for example the US, QRA is more widely used as a means of assessing risk. If one knew that QRA produced dependable results then clearly one could adopt the process as a means of standard setting. Given that some level of acceptable risk could be defined, then a simple calculation would lead one to the acceptable level of exposure of the compound in question. The World Health Organization (WHO) in its Air Quality Guidelines for Europe adopted the techniques of QRA but did not carry the process forward to guide-line-setting. It was pointed out by WHO that the Air Quality Guidelines were designed to indicate concentrations of compounds on exposure to which the great majority of the population, including sensitive groups, would experience no ill effects. Given that for many genotoxic carcinogens toxicologists agree that no completely safe level of exposure can be assumed, WHO felt that recommending guidelines for such compounds would be unsatisfactory.

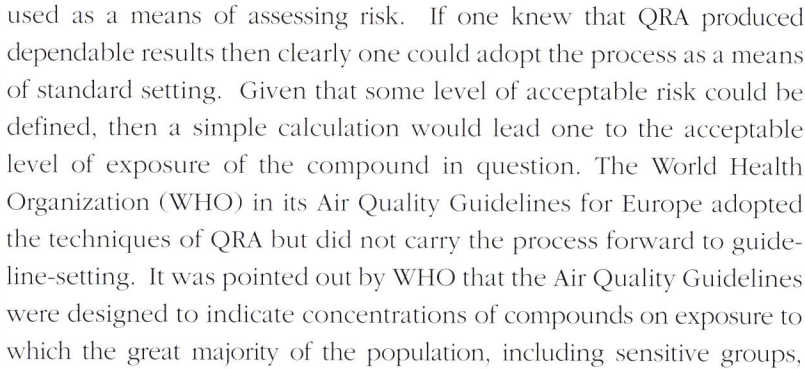

EPAQS reports on carcinogens

3.8   In the UK the Department of the Environment's Expert Panel on Air Quality Standards (EPAQS) has examined two carcinogenic compounds, benzene and 1,3-butadiene and has recommended air quality standards. The Panel did not use a QRA approach but, instead, identified from studies of the effects of occupational exposure, levels of exposure which had not been demonstrated to be associated with increased risk compared with background levels in unexposed populations. It was acknowledged that such levels could not be accepted as definitive No Observed Adverse Effects Level (NOAELs) as large studies might have detected effects at lower levels of exposure. However, by using such values as a starting point and by applying a series of safety factors, standards were derived.

3.9   One safety factor was used to allow for the fact that exposure in the occupational setting is likely to be for not more than 8 hours per day for the working life in comparison with a 24 hours per day, 70 year, lifetime, exposure. A second safety factor was used to allow for the existence in the whole population of individuals who might be significantly more sensitive to the compound in question than the work-forces studied. A value of 10 was assigned to each safety factor. This is in line with standard toxicological practice and seems to have provided an adequate margin of safety during the forty or so years that such factors have been in use.

3.10   In the case of 1,3-butadiene a further point was taken into account. It is desirable that people's exposure to carcinogens should not be allowed to increase. This is especially the case with regard to that group of carcinogens which are described as genotoxic and for which it is plausible that no absolutely safe level of exposure may exist. Thus the caveat, ambient levels should not be allowed to rise, was noted and applied.

3.11 This process allowed the Panel to recommend air quality standards which they felt would provide adequate protection of health. The Panel concluded, for example, in the case of benzene that exposure to the concentration of benzene defined by the standard (5 ppb, annual average concentration) would "present(s) a risk to the population of the United Kingdom which is exceedingly small and unlikely to be detectable by any practicable method." It should be noted that it was not asserted that no risk would be presented, only that, in the judgement of the Panel, the risk would be exceedingly small.

Setting standards for non-carcinogens

3.12 For compounds not likely to produce cancer the process of standard setting is better established. Studies of the effects of these compounds on volunteers and epidemiological studies of the effects of exposure of populations allow identification of levels of exposure at which no effects occur. Of course, exposure to any compound may include minor effects which might not be regarded as adverse. For example, a transient, small change in some index of lung function might not be thought to be of concern. Because of this, identification of a No Observed Adverse Effect Level (NOAEL) has been preferred in standard setting. Sometimes such a level of exposure has not been identified and only a Lowest Observed Adverse Effect Level (LOAEL) has been defined.

3.13 It is accepted that chamber studies, involving the exposure of volunteers to pollutants, whilst providing a controlled environment and the possibility of undertaking detailed physiological measurements, do not fully reflect the effects of exposure to ambient pollutants. Chamber studies tend to be limited in duration, to involve the participation of healthy, or at least not severely ill subjects, and in general, the use of single pollutants. Chamber studies do, however, allow the detailed level of study which is necessary for understanding the mechanism of effect of air pollutants. Studies in experimental animals are also important in contributing in this way. Recent chamber studies have involved exposure of volunteers to combinations of allergens and pollutants. Such studies have shed light on important interactions and have helped to explain the results of some epidemiological work.

3.14 Epidemiological studies of air pollutants have a long history. Studies of air pollution episodes, of the effects of day to day changes in levels of air pollution on health and of the effects of long term exposure to pollution have

---

Concentration - Exposure - Dose

Knowing the concentration of a pollutant alone does not allow prediction of its effects. For there to be effects people have to be exposed to the pollutant, often the effects depend on the duration of exposure. The actual dose received depends additionally on physical activity: the harder you exercise the more pollutant you inhale and thus the greater the possible effects.

Somebody who jogs for long periods in the park on a day when ozone levels are moderate may in fact receive a much greater dose of ozone than somebody else who stays indoors on a day of exceptionally high ozone concentrations.

all played important roles. All such studies take advantage of variations in exposure of individuals to air pollutants. It is therefore necessary to characterise exposure as well as possible. This is far from easy and fixed site, outdoor monitoring of air pollutants provides only a relatively coarse estimate of individual exposure. In many developed countries, such as the UK, it is known that individuals spend more than 80% of their time indoors. Levels of pollutants indoors may differ considerably from those outdoors, and the indoor/outdoor relationship may differ significantly from pollutant to pollutant. For example, ozone reacts with many materials found indoors and the indoor concentration is generally very considerably less than that outdoors. On the other hand, in many houses, fine particulate material may penetrate well into the indoor environment and indoor concentrations may follow and approximate to those outdoors. Indoor sources of fine particles also exist. Careful studies of personal exposure are needed if the effects of pollutants, at the level of the individual, are to be understood.

3.15   Fortunately, by adopting a cautious approach and applying safety factors to take account of these uncertainties, air quality standards can be set for pollutants outdoors. The magnitude of the safety factors adopted frequently depends on the quality of the data available. Ozone provides a good example.

3.16   Ozone is the most potent irritative or pro-inflammatory pollutant of the common range of air pollutants. Chamber studies have shown that exposure to concentrations of the order of 80 ppb for about six hours in subjects undertaking intermittent exercise leads to an increase in airway resistance as an index of airway narrowing and the release of inflammatory mediators into the respiratory tract. If the only data available related to healthy adult volunteers then we might feel that the application of a substantial safety factor would be necessary in defining a standard designed to protect all the population including children suffering from asthma who might be exposed for long periods whilst playing outdoors on warm summer days.

3.17   Consideration of safety factors has been helped by studies of children with asthma at summer camp in the United States, which allowed an accurate estimate of the effects of ozone on such children. With this knowledge the need for a large safety factor is obviated and the standard may be set closer to the LOAEL value.

3.18   In setting standards for air pollutants many groups, including EPAQS, have adopted a "weight of evidence" approach. In any series of studies of the effects of a given exposure to a pollutant identical results from study to study are unlikely. Of course, the results tend to cluster and from this some estimate of the expected effect is made. Outlier results which differ substantially from those of other studies need very careful consideration. The usual scientific test of repeatability by the original author and by other workers is applied to such results before they are given weight in the standard setting process. It is clear that an important element of expert judgement is needed here and experienced clinicians and toxicologists have been appointed to EPAQS to ensure that such is available.

Sulphur dioxide

3.19 Sulphur dioxide (SO_2) is the principal pollutant associated with acid deposition, usually after oxidation to sulphuric acid. Sulphuric acid is generally present as an acid aerosol, often associated with other pollutants in droplets or solid particles of various sizes.

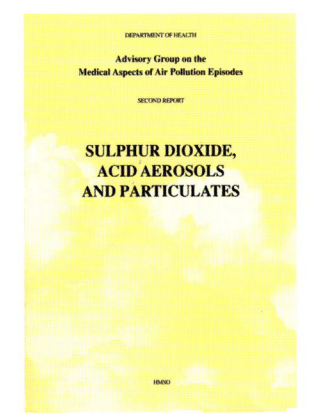

DEPARTMENT OF HEALTH

Advisory Group on the
Medical Aspects of Air Pollution Episodes

SECOND REPORT

**SULPHUR DIOXIDE,
ACID AEROSOLS
AND PARTICULATES**

HMSO

3.20 Concentrations of sulphur dioxide have declined during the past fifty years in many European countries. This is due to the general move away from the use of coal as a means of domestic heating. High concentrations do, however, occur in Eastern European countries. Exceptional concentrations may occur on a very local basis if a plume of smoke from a chimney (eg an industrial chimney) falls to the ground due to local atmospheric conditions. Such events are described as "fumigations".

3.21 In the period before smoke control in UK cities, effects on health were demonstrated in a series of epidemiological studies. In this period, high concentrations of sulphur dioxide were commonly found together with much higher concentrations of smoke than seen today. Daily death rates and worsening of the condition of chronic bronchitics, appeared to be related to raised levels of smoke and SO_2.

3.22 In the UK, SO_2 still arises principally as a result of the combustion of fossil fuels. Non-nuclear power stations account for 66% of total UK emissions and other industrial combustion 16% (1993 figures). Emissions declined by 50% between 1970 and 1993. Vehicles are not a significant source of SO_2 (2%), although diesel engines produce more than petrol engines.

3.23 The studies in London in the 1950s and 60s formed an important part of the data base upon which the WHO Air Quality Guidelines for Europe were established in 1987: a 24 hour guideline of 125 $\mu g/m^3$ for SO_2 and the same concentration for black smoke was recommended. In addition, an equivalent value of 70 $\mu g/m^3$ for Thoracic Particles (an approximate equivalent of PM_{10}) was recommended.

In 1994 the WHO Air Quality Guidelines for sulphur dioxide and particles were reviewed by a WHO expert panel and a recommendation was made to separate the guideline for sulphur dioxide from that for particles. The 24 hour guideline for SO_2 was retained as 125 $\mu g/m^3$. This represented an important change in policy. If the possible health effects of pollution mixtures similar to those found in London forty years ago are being considered the potentially important interactions of sulphur dioxide and smoke to increase levels of airborne acidity should still be borne in mind.

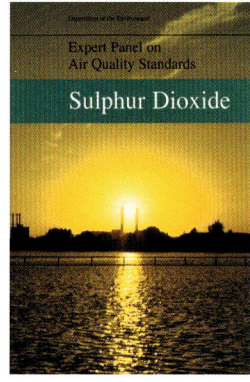

Standards and guidelines (SO$_2$) (WHO Air Quality Guideline values and EC Directive Limit Values for SO$_2$ can be found in Appendix 1.)

DoE banding: (peak hourly average in 24 hour period)	
0 - 59 ppb	very good
60 - 124 ppb	good
125 - 399 ppb	poor
> 400 ppb	very poor

National Air Quality Standard:

15 min: 100 ppb

Levels in UK

3.24   High concentrations of SO$_2$ (daily average on peak days of > 1300 ppb) together with high concentrations of suspended particles are believed to have been responsible for the high mortality levels during the London smog of 1952.

3.25   In urban areas SO$_2$ levels have shown a marked decline throughout the 1960s and 1970s. Urban and rural levels are now similar, as are summer and winter levels, showing the reduction in the domestic use of coal and fuel oil. In Belfast, however, coal and solid smokeless fuel are still widely used (in the absence of piped natural gas), and there are very significant peak levels (of the order of 500 ppb hourly average) of SO$_2$ in the winter.

Fig 3.2
Long-term trends in annual mean concentrations of smoke and sulphur dioxide in Greater London

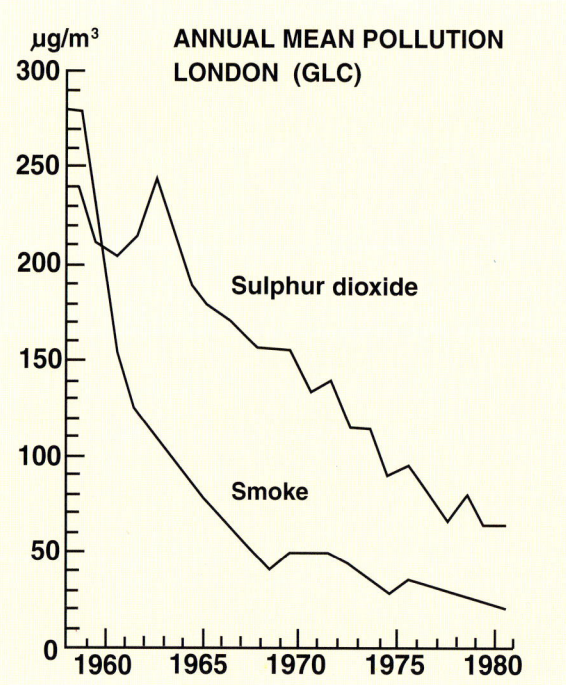

3.26   Annual mean concentrations in urban areas are generally in the range 10 - 20 ppb, but maximum one hour means can be as high as 200 - 300 ppb at most mainland monitoring sites, although 1 hour maximum levels as high as 544 ppb have been recorded in Belfast in recent years.

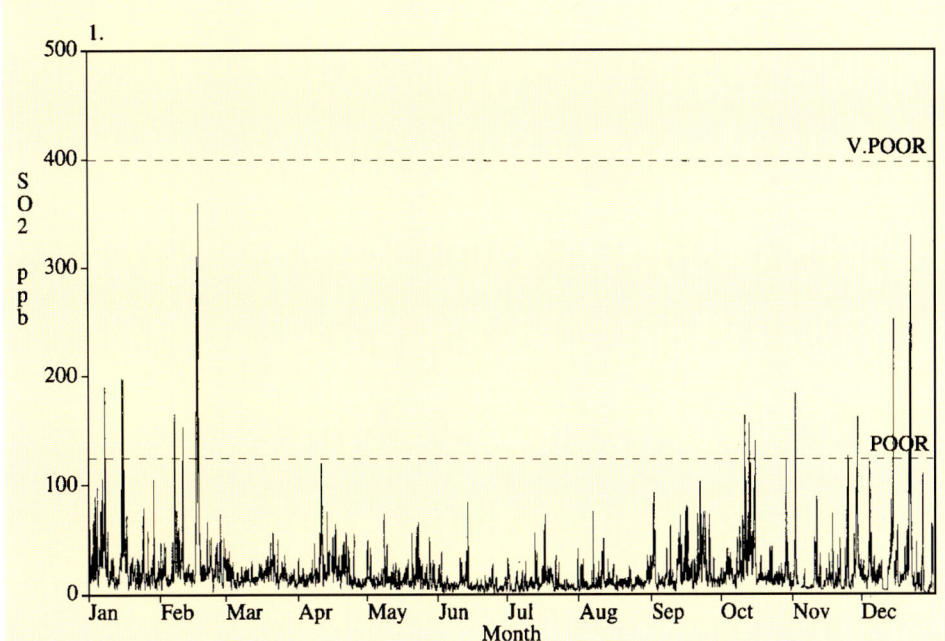

Fig 3.2
Belfast East Sulphur
Dioxide 1994

Health effects

3.27   SO$_2$ concentrations of more than 4000 ppb can have severe effects and concentrations of more than 1000 ppb can give rise to immediate problems, such as bronchoconstriction (narrowing of the airways), for asthmatics. Studies indicate reversible changes in child lung function at 100-150 ppb, aggravation of bronchitis (above 200 ppb) and perhaps an increase in mortality above 200-400 ppb.

3.28   Though the effects of sulphur dioxide on public health have been studied intensively for more than forty years, the exact mechanism of effect, beyond its capacity to excite sensory nerve endings in the lung, remains obscure. It is well known that inhalation of sulphur dioxide produces bronchoconstriction and that asthmatic patients are more sensitive than other individuals. Small effects on standard measures of lung function may be expected on exposure of asthmatic subjects for just a few minutes at concentrations of sulphur dioxide as low as 200 ppb (572 µg/m^3). Such concentrations may occur during fumigation. Because of these effects of short term exposures, guidelines and standards for SO$_2$ usually include a short-term (10 or 15 minutes) value.

3.29   Air pollution episodes with sharp increases in smoke and sulphur dioxide have been associated with acute effects on health. The best known example was the London fog in December 1952 which lasted for 5 days. The concentrations of SO$_2$ (greater than 1,300 ppb/3,800 µg/m^3) and smoke (around 4,400 µg/m^3) were around 100 times higher than typical urban values today. There were an estimated 4,000 additional deaths, mainly amongst the elderly and chronically ill, and a steep rise in respiratory ailments.

In these studies it was not possible to determine whether the smoke or the SO_2 was having the greater effect and the possibility that the effects were due to acid being transported on the surface of the smoke particles was raised. Re-analysis of data collected in London has supported the concept that aerosol acidity is of major importance in producing effects on health.

3.30   Recent epidemiological studies conducted in countries not known for high concentrations of sulphur dioxide, including Switzerland, have demonstrated that lung function may be related to long term average concentrations of sulphur dioxide. For example, a study compared the lung function of residents of different regions of Switzerland with annual SO_2 concentrations varying from 1-9 ppb (2-26 µg/m^3). The results suggested that for every 10 µg/m^3 there was a decrease in lung function in healthy non-smokers of about 3%. A 10% increase in wheeze and a 20% increase in breathlessness during exercise and chronic cough was also estimated for the same increase in concentration. The mechanism of this long term effect is not understood but the studies should not be ignored. In 1994, concern generated by these studies and the results of the older studies led the WHO to confirm its guideline of 17 ppb (50 µg/m^3) as an annual average concentration. It is not possible to determine whether the relationship demonstrated between long term average concentrations of SO_2 and lung function is causally dependent on exposure to SO_2. This is a problem of many epidemiological studies. It is possible that SO_2 acts as a surrogate for other pollutants, perhaps most likely for fine acidic particles.

Nitrogen dioxide

3.31   The oxides of nitrogen (referred to collectively as NO_x) comprise several gases, including nitric oxide (NO) and nitrogen dioxide (NO_2). In the ambient air, NO_2 is probably the most important for human health, so that data on health risks and guidelines are usually expressed in terms of NO_2 rather than NO_x.

Sources

3.32   Nitrogen dioxide is produced both directly as a primary and indirectly as a secondary pollutant owing to the spontaneous conversion of NO to NO_2 in the presence of ozone or oxygen. In the UK, some 50% of the atmospheric nitrogen dioxide is produced by motor vehicles and 25% by power stations. Although it is mainly a secondary air pollutant, NO_2 is rapidly formed close to the sources of NO. In busy streets, where concentrations tend to be high, the ratio of NO_2:NO is about 4:1. In rural areas where more of the NO has been converted to NO_2 the ratio may be 9:1. NO_2 is also a common indoor air pollutant, being generated by gas cookers and heaters. NO reacts rapidly with ozone to form NO_2 and this leads to a depletion of ozone close to busy roads.

Outdoor concentrations

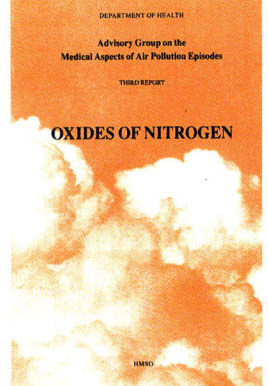

3.33   Concentrations of NO_2 in urban areas show a distinct diurnal variation with peak levels typically being recorded during the morning and evening rush hours. In cold, still weather, concentrations may increase due to trapping of pollutants in a layer of cold air close to the ground. These episodic increases in concentrations of NO_2 do not reflect a change in sources but a change in the pattern of dispersion of the pollutant. In London, in December 1991

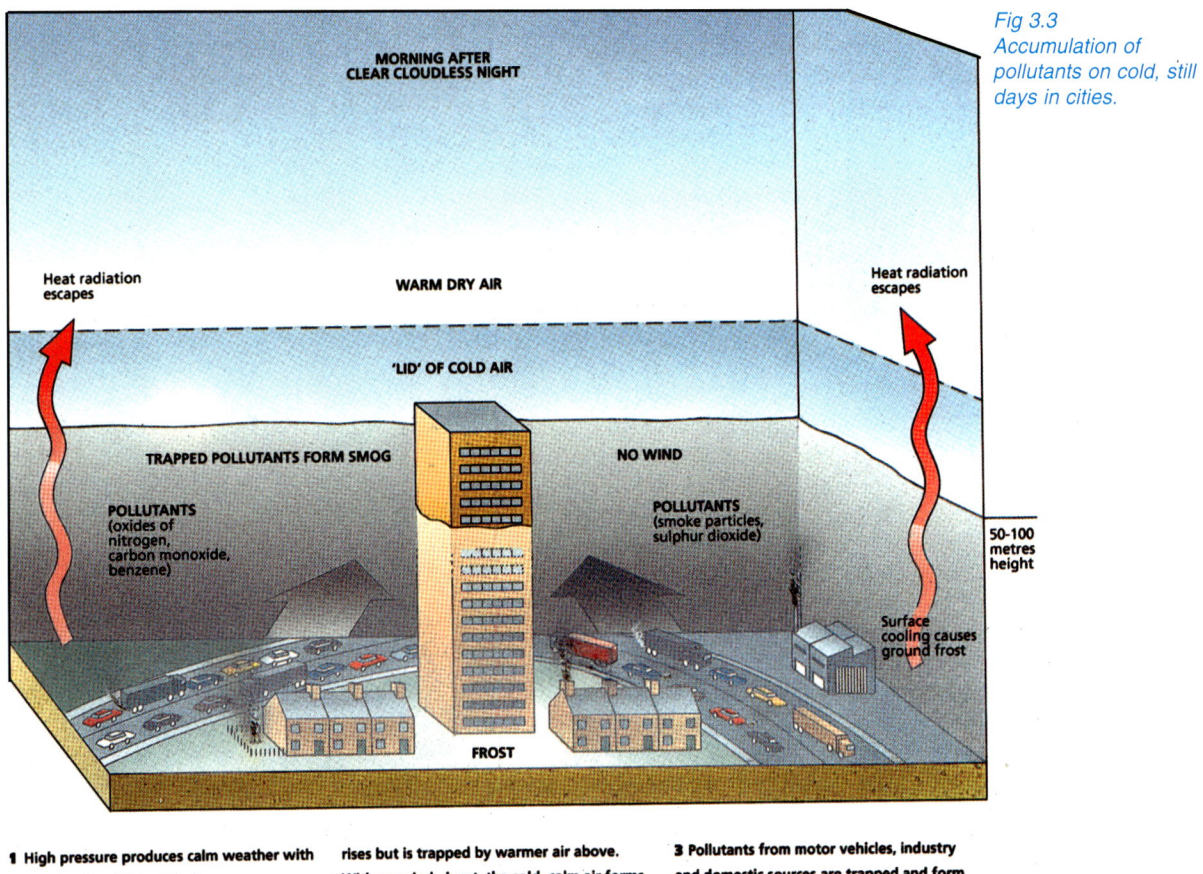

Fig 3.3
Accumulation of pollutants on cold, still days in cities.

1 High pressure produces calm weather with warm dry air at high altitudes.
2 Ground level temperatures fall and cold air rises but is trapped by warmer air above. With no wind about, the cold, calm air forms a layer or 'lid'.
3 Pollutants from motor vehicles, industry and domestic sources are trapped and form ground-level smog.

concentrations of NO_2 across the city rose to more than 400 ppb (752 µg/m³). It follows that the atmospheric conditions which cause an increase in concentrations of NO_2 also cause an increase in concentrations of other, especially primary, pollutant concentrations. Particles and carbon monoxide, for example, would be expected to rise in parallel with those of NO_2. The hourly concentrations of NO_2 in London during 1991 are shown in Figure 3.5 the December episode is obvious.

3.34 NO_2 levels in the UK are generally higher in urban areas (particularly London) due to levels of traffic. Annual mean concentrations in urban areas are generally in the range 10-45 ppb, whilst maximum daily and one hour means can exceed 200 ppb and exceptionally 450 ppb.

3.35   During an episode in London in December 1991, levels in central London reached 423 ppb. Similar episodes occurred in Manchester and Walsall in December 1992 (peaks 369 and 269 ppb respectively). An episode moved southwards across the country during 22 - 23 December 1994, with a peak hourly average of 352 ppb in Manchester at 1200 on 22/12, a peak hourly average of 257 ppb in Walsall at 0700 on 23/12 and a peak hourly average in London of 268 ppb during the evening rush hour of 23/12.

Fig 3.4
Hourly concentrations of NO$_2$, at
Bridge Place, London 1991-1992

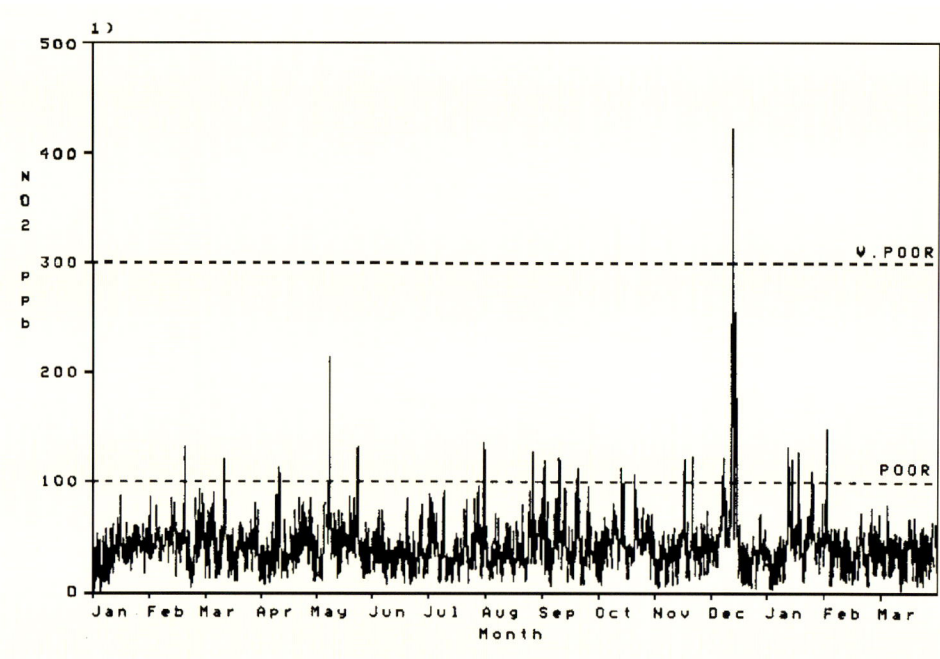

Indoor concentrations

3.36   Levels of nitrogen dioxide in indoor air are largely determined by the level outdoors unless there is an indoor source of the pollutant. The most significant indoor source of NO$_2$ is cooking with gas, although other unflued gas appliances may also contribute. About half the homes in Britain have gas cookers.

Fig 3.5
Hourly concentrations of NO$_2$ at
Bridge Place (London) 1994.

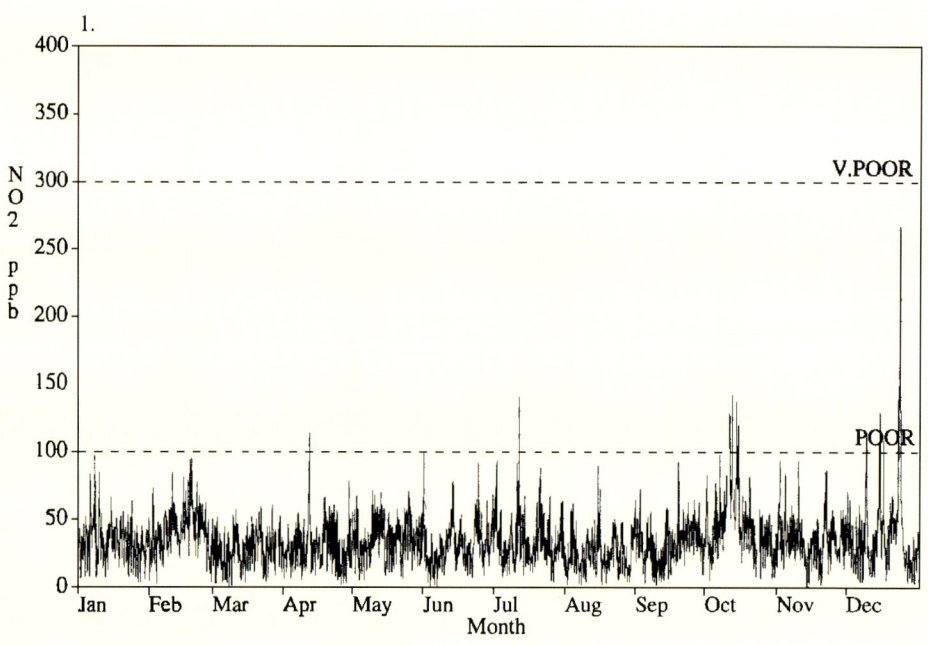

3.37   Averaged over one or two weeks, levels of NO$_2$ have been found to range from 25-70 µg/m^3 in homes where there is gas cooking. One hour average concentrations can reach 1115 µg/m^3. Hence it is possible that in using gas for cooking, NO$_2$ levels in the home, particularly in the kitchen, could exceed the WHO guideline value. By contrast, homes without gas cookers have average weekly levels in the region of 13-40 µg/m^3.

3.38 The combustion of gas produces large amounts of water vapour and carbon dioxide, together with traces of other compounds (such as carbon monoxide, formaldehyde and particles), all of which may confound any analysis of the health effects of NO_2 derived from gas cookers.

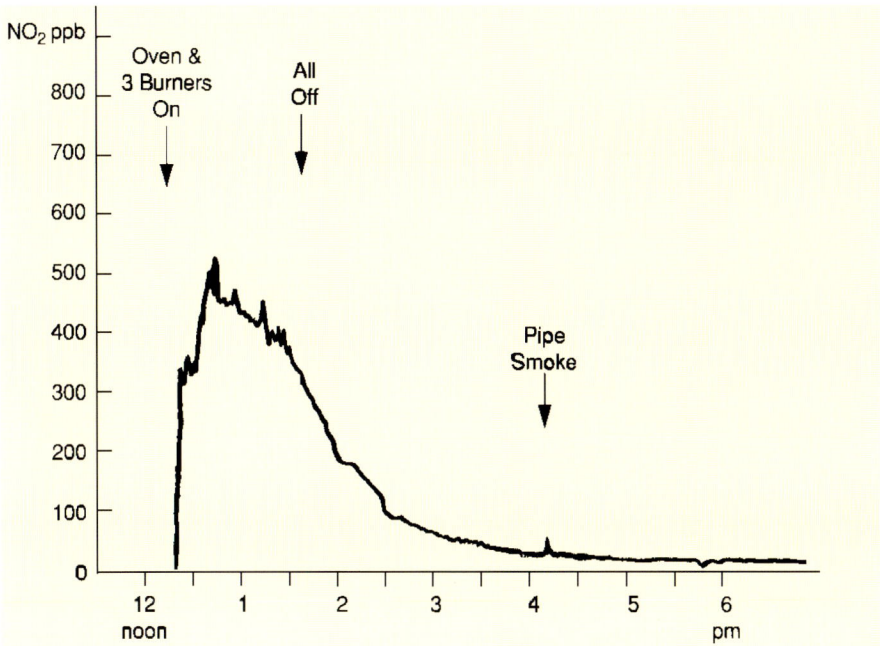

Fig 3.6
Example of variation of nitrogen dioxide concentration in a domestic kitchen following the use of a gas cooker. R E Waller, personal communication

Health effects of NO_2

(a) Experimental evidence

3.39 Nitrogen dioxide is an irritant of the airways and exposure to concentrations far above those encountered either outdoors or indoors in the UK can produce bronchoconstriction in both asthmatic and non-asthmatic individuals. Exposure to concentrations of NO_2 of about 300 ppb (560 μg/m^3) for 30 minutes produces a small change in measures of lung function in asthmatic individuals. In non-asthmatics exposure to about 1 ppm (1800 μg/m^3) is necessary to produce a similar response. The exposure response relationship for NO_2 is both unusual and interesting. Exposure to concentrations of 300 ppb may produce a response whilst exposure to 600 ppb may not. The response seems to reappear and to remain as concentrations approach and exceed 1000 ppb (1 ppm). The explanation for this unorthodox exposure-response relationship is obscure.

3.40 Exposure to NO_2 at concentrations not very much in excess of those occasionally experienced both indoors and out has been shown to increase the response of sensitive individuals to allergens. The importance of this effect on public health is uncertain.

(b) Effects of outdoor sources

3.23 Young children and subjects with asthma have been seen as sensitive groups who may be affected during NO_2 pollution episodes. Individuals with chronic bronchitis, emphysema or other chronic respiratory diseases may also be sensitive to NO_2 exposure. Epidemiological evidence of effects of nitrogen dioxide in

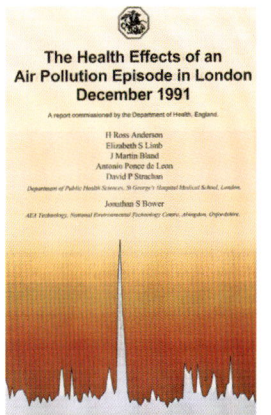

Report on the London 1991 episode commissioned by DH

asthmatics is more mixed but increased sensitivity amongst asthmatics has not been conclusively demonstrated. Some studies have shown greater sensitivity; others have not. There is some evidence to suggest that exposure to relatively high ambient concentrations of NO_2 (reached occasionally in the UK) can enhance the response to allergens and substances which produce constriction of the airways in asthmatics. At low levels of exposure the evidence is inconsistent. Some, but not all, studies have suggested an increased prevalence of asthma in those living close to busy roads. Some believe this may be due to the increased concentrations of nitrogen dioxide near roads.

3.41   During the episode of pollution that occurred in London in December 1991 there was a rise in the death rate and in admissions to hospital, suggesting that severe pollution can have serious effects, particularly on susceptible people. Levels of nitrogen dioxide were very high but particle concentrations were also raised. It was thus difficult to distinguish whether nitrogen dioxide or particles were responsible.

(c)   Effects of indoor sources

3.42   Although the evidence is not wholly consistent, several studies have shown that children living in houses containing gas cookers have a higher risk of respiratory infections than children in houses with electric cookers. When the results of these studies are combined, it seems that children aged 5-12 years have a 20% increase in risk of respiratory infection for each increase of 15 ppb ($28.3 \ \mu g/m^3$) in the two-week average concentration of NO_2 across a range of weekly average concentrations of 8-65 ppb ($15-128 \ \mu g/m^3$). This effect has been attributed to long term exposure to raised concentrations of nitrogen dioxide. Careful examination of the data has shown that the relationship between infection rates and the use of gas cookers is stronger than the relationship between infection rates and concentrations of NO_2. That NO_2 is the causative agent of the effect is likely but not yet certain. A single study in the UK has shown that long term exposure to the pollutants generated by gas cooking leads to a depression of standard indices of lung function amongst women. This relationship has been attributed to the effects of NO_2 exposure, but conclusions cannot be drawn with certainty.

3.43   The risk of respiratory illness from the levels of NO_2 currently found in most homes appears to be small. Nevertheless, measures to reduce domestic exposure to NO_2 and other combustion products are being encouraged because of the large number of people potentially exposed and because of the uncertainties regarding the effects on susceptible groups such as asthmatics and bronchitics.

Standards and guidelines (NO$_2$) (WHO Air Quality Guideline Values and EC Directive Limit Values can be found in Appendix 1.)

DoE banding: (peak hourly average in a 24 hr period)	
0 - 49 ppb	very good
50 - 99 ppb	good
100- 299 ppb	poor
> 300 ppb	very poor

National Air Quality Standard:
1 hour average concentration: 150 ppb
Annual average concentration: 21ppb

3.44   MAAPE's third report on Oxides of Nitrogen (1993) concluded that:

"The available evidence indicates that individuals not suffering from respiratory disease will be unaffected by such episodes of elevated concentrations of nitrogen dioxide as occur in the UK. When studied in the laboratory there is no consistent difference in sensitivity to nitrogen dioxide between asthmatic patients and normal individuals. However, some recent epidemiological studies have indicated that people suffering from respiratory disorders, including asthma, may experience a worsening of their symptoms when ambient levels of nitrogen dioxide and associated pollutants are raised".

Ozone

3.45   Ozone (O$_3$) is a highly reactive oxidising agent. In the lower atmosphere, tropospheric ozone is formed indirectly by the action of sunlight on NO$_2$. The rate of ozone production depends on the concentrations of the reactive compounds and the intensity of sunlight. Volatile organic compounds (VOCs) contribute substantially to atmospheric photochemical reactions and thus ozone formation. The lifetime of ozone in polluted areas is about one day, much is removed indoors and at night by reaction with surfaces.

Sources

3.46   Ozone is a secondary pollutant. There is evidence that, compared with the preindustrial period, background level and concentrations have doubled over the last 100 years. Episodes in which levels rise substantially above background occur in summer when there are long hours of bright sunlight, temperatures above 20°C and light winds. Ozone tends to build up downwind of urban conurbations (where most oxides of nitrogen are emitted by road transport). Once formed, ozone can travel large distances and peak levels tend to occur in rural areas.

In urban areas levels tend to be lower because of the scavenging effect of urban levels of NO and the time taken for O_3 to be produced from its precursors. Significantly raised levels can, however, occur in suburban areas, where population density is high. Ozone is a transboundary pollutant and in summer a significant proportion of ozone in Southern England is formed from primary pollutants emitted in Europe.

Standards and guidelines (O_3): (WHO Air Quality Guideline Values and EC Directive Values can be found in Appendix 1.)

NB: The following standards and guidelines are based on either 1 hour or 8 hour averaging of ozone concentrations. EPAQS and the most recent WHO recommendations (see Appendix 1) took the view that an 8 hour averaging time was the most appropriate; however earlier guidelines use a 1 hour average. It is important to note the averaging time: 1 hour and 8 hour averages cannot be regarded as equivalent as the response to ozone is time dependent.

DoE banding: (peak 1 hr average in a 24 hour period)	
0 - 49 ppb	very good
50 - 89 ppb	good
90 - 179 ppb	poor
> 180 ppb	very poor

National Air Quality Standard:
Moving 8 hr average: 50 ppb

UK levels

3.47   The highest recorded level in the UK was 258 ppb at Harwell in 1976, when levels in London regularly exceeded 200 ppb during a two-week period. Levels exceeding 180 ppb were not recorded in the UK between 1981 and 1994, and during the hot, sunny weather of June/July 1994 100 ppb was exceeded only occasionally (a maximum of 4 days at any one site and a maximum concentration of 119 ppb) at rural monitoring sites in Southern England. There were prolonged periods of high temperatures during the summer of 1995: the peak 1 hour concentration was 134 ppb. The reasons for the lower than expected concentrations experienced during 1995 are currently being investigated by the DoE Photochemical Oxidants Review Group.

3.48   Long term trends in ozone levels in the UK show a steady though slight decline in urban areas (probably due to increasing local NO emissions), while there is an upward trend (~ 1 ppb per year) in rural sites in the north of the country. The overall pattern seems to be that concentrations are declining slowly in the south and south-east, but increasing (consistent with global trends) elsewhere.

Health effects

3.49   Ozone is a very active oxidising agent which attacks biological materials including cell membranes and proteins. In terms of producing inflammation of the respiratory tract ozone is the most toxic of the common air pollutants. Exposure for about six hours to concentrations as low as 80 ppb (160 µg/m³) produces both inflammation of the airways and changes in standard indices of lung function.

3.50   Its reactivity is such that concentrations indoors are rapidly reduced by reaction with plastics and fabrics and, almost alone amongst the major air pollutants, exposure to ozone is solely an outdoor-air problem.

3.51   Ozone and associated photochemically-produced pollutants at hourly levels in excess of 100 ppb cause eye, nose and throat irritation, chest discomfort and cough, probably through irritation of airway nerves. Chamber studies have revealed that about 10% of the population has an increased sensitivity to ozone. Individuals not in this group are unlikely to experience effects at UK ambient concentrations.

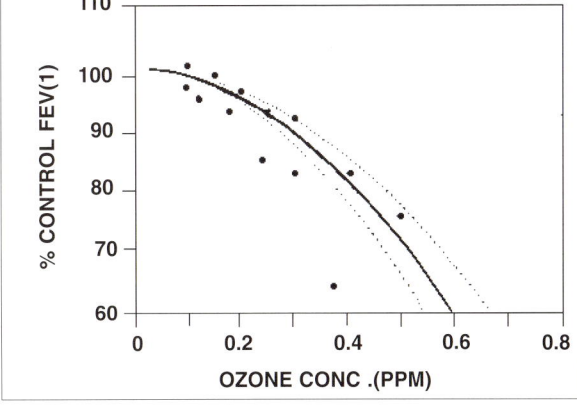

Fig 3.7
Scatter plot of O₃, concentration vs. forced expiratory volume at 1 second during very severe exercise. Regression line and 95% confidence limits are superimposed.

3.52   The biological response to ozone is dependent upon the concentration to which an individual is exposed and the duration of exposure. In addition, the dose received will be dependent upon the volume of air inhaled per minute as nearly all the ozone inhaled is absorbed. Long duration (6.6 hr) experimental studies in exercising subjects, have shown that the decline in indices of lung function is proportional to the physical work rate of the subjects. The effect of exposure to a range of concentrations is shown in Fig 3.7. The importance of duration of exposure and the appreciation that this may extend over 8 hours on a sunny day, has led to guidelines and air quality standards for ozone being typically defined in terms of an eight hour average concentration.

3.53   Ozone produces constriction of the airways. It might therefore be expected that individuals who suffer from asthma would be significantly more responsive to ozone than other individuals. This has, however, not turned out to be true, though there is limited evidence to suggest that asthmatics produce a more marked and less rapidly resolving inflammatory response to ozone than other individuals. More important than this is the inter-subject variation in response to ozone. This is substantial. Exposure to ozone which produces on average only a few percent change in FEV_1, (a standard index of lung function) may in some subjects produce a change of 30%. Should such a change be superimposed upon already depressed lung function then symptoms may be expected. Because of this, some asthmatics may be particularly vulnerable to ozone.

3.54 Epidemiological studies have confirmed this view: studies in Canada and the US have produced compelling evidence that emergency admissions of asthmatics to hospital increase on days when concentrations of ozone are raised.

3.55 The biological response to ozone is in one way, however, unusual though not unique. If exposure is continued day to day the response falls away and the subject is described as having developed tolerance to ozone. This phenomenon has not been adequately explained and two contrasting explanations are possible:

a. the subject has become so damaged that no further response is possible.

b. the subject has increased his defence mechanisms, for example, the concentration of natural anti-oxidants in the respiratory tract, and is thus no longer affected.

Each explanation is plausible. The waning of the response should not be taken as an indication that day after day exposure does no more harm than a single day's exposure.

3.56 In 1994 a WHO Expert Group which reviewed the Air Quality Guideline for ozone took into account recent outdoor studies undertaken in the US which have shown that exposure to 60 ppb would not be expected to produce significantly adverse effects. In addition to revising the Guideline, two tables producing guidance on the health outcomes likely to be associated with exposure to different concentrations of ozone were also provided. See tables 3.1 and 3.2.

Table 3.1
Concentration - response
relationships for Ozone:
WHO 1994.

Health outcomes associated with changes in ambient concentration in epidemiological studies	Change in 1h O_3 ($\mu g/m^3$)	Change in 8h O_3 ($\mu g/m^3$)
Symptom exacerbations among healthy adults or asthmatics - normal activity		
25% increase	200	100
50% increase	400	200
100% increase	800	300
Hospital admissions for respiratory conditions		
5% increase	30	25
10% increase	60	50
20% increase	120	100

Health outcomes associated with controlled ozone exposures		O_3 concentration (µg/m³) at which health effect expected	
	Averaging	1 h O_3	8 h O_3
FEV$_1$ (active, healthy, outdoors, most sensitive 10% of healthy children & young adults			
5%		250	120
10%		350	160
20%		500	240
Inflammatory changes (neutrophil influx) (healthy young adults at >40 1/min outdoors)			
2-fold increase		400	180
4-fold increase		600	250
8-fold increase		800	320

3.57   Caution must be exercised in applying the results shown in these tables. It will be seen that the relationship defined in the tables is linear in terms of concentration of ozone and response. Uncertainty, however, exists about the form of this relationship, the curve becoming more steep as the concentration increases. Abatement strategies might therefore be expected to produce greater effects per unit fall in ozone concentration where levels are greatest. In addition, some uncertainty exists regarding the effects of co-pollutants on the response to ozone. In some studies the co-existence of an acidic aerosol may have contributed to the measured response.

3.58   One interesting effect of ozone not discussed above is its effect on the response of the airways of sensitised subjects to allergen. Ozone has been demonstrated to increase the magnitude or sensitivity of the airways response to inhaled pollen allergen in sensitised individuals. This has been confirmed in several volunteer studies and in an epidemiological study of hay-fever sufferers in London. The co-occurrence of peak levels of pollen and of ozone during the summer may make this an important observation.

3.59   The main conclusion of the MAAPE report in 1991 was that:

> "Changes in lung function may occur in people during such episodes of elevated ozone concentration as are found in parts of the UK during periods of hot summer weather. These changes are unlikely to produce irreversible lung damage though individuals who are sensitive to ozone may experience respiratory symptoms, including cough and discomfort on deep inspiration, whilst taking vigorous exercise out of doors. Although individuals with asthma or other respiratory disorders appear to be no more likely than healthy individuals to be sensitive to ozone, the effects of ozone may be more troublesome in individuals who already have some impairment of lung function".

3.60   Over the past few years newly emerging evidence that asthmatics may in fact be a little more sensitive to ozone than are normal individuals has been accumulating. Though asthmatic subjects do not seem to be characterised by greater bronchoconstrictor responses to exposure to ozone than non-asthmatic people, the inflammatory response in the airways is greater and longer lasting.

3.61   This was discussed further at COMEAP in 1993 and the suggestion that the advice on the DoE Air Quality Helpline might be changed was considered. The view of the Committee was that there was insufficient evidence to require that asthmatics be provided with more advice. The importance of encouraging asthmatics to live as normal a life as possible was considered as most important by the Committee and it was felt that this approach might be undermined by providing more urgent warnings when on the basis of the evidence available risks were judged to be small.

Particulate matter

3.62   This is a complex mixture of organic and inorganic substances. Fine particles (< 2.5 µm) comprise secondarily formed aerosols, combustion particles and recondensed organic and metallic vapours.

Sources

3.63   Particles can be primary (emitted directly to atmosphere) or secondary (for example, formed as sulphates and nitrates from other pollutants such as SO_2 or NO_2). Primary particles are directly emitted from sources such as non-nuclear power stations, motor vehicles, cement factories and open cast coal mines. They also occur naturally as airborne spores and pollen grains and their fragments. The most abundant secondary constituent is frequently ammonium sulphate, formed from ammonia and sulphuric acid, and found in both urban and rural air.

3.64   In urban areas motor vehicles are the major source of particulates, with diesel vehicles being responsible for most of the black smoke. In London 85% of particles measured as black smoke are thought to be produced by diesel powered vehicles. There is some uncertainty regarding the total vehicular contribution to urban particulate concentrations measured as PM_{10}; estimates range from 20-50%. Domestic coal burning remains an important source in Belfast and also in some towns in coal mining areas.

Terms and units of measurement

All these descriptors are expressed as mass in $µg/m^3$

Inhalable dust: <15 µm aerodynamic diameter

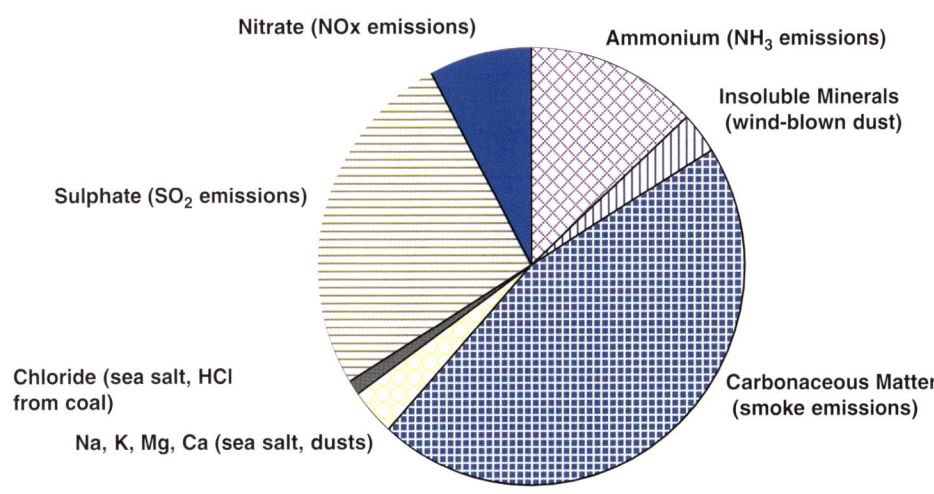

Ammonium (NH₃ emissions)

Insoluble Minerals
(wind-blown dust)

Sulphate (SO₂ emissions)

Chloride (sea salt, HCl
from coal)

Na, K, Mg, Ca (sea salt, dusts)

Carbonaceous Matter
(smoke emissions)

Fig 3.8
Composition and major source
categories for atmospheric particles
sampled in Leeds

Respirable dust:	<5 μm (MRC definition)
Fine particles:	<2.5 μm
Ultrafine particles:	<0.1 μm

PM_{10} (also thoracic particles [TP]): <10 μm (more strictly, all particles passing through a size selective inlet which allows through 50% of 10 μm particles. The selection curve is steep at 10 μm, so PM_{10} is essentially equivalent to all particles <10 μm)

$PM_{2.5}$: similar to PM_{10}, but <2.5 μm

Suspended particulate matter (SPM): a general term embracing all airborne particles.

Thoracic particles (TP): see PM_{10}

Total suspended particulates (TSP): mass of airborne particles

Smoke: particles <15 μm, derived from incomplete combustion of fuels.

Black smoke (BS): non-reflective particulate matter associated with measurement of reflectance of a stain produced by drawing air through filter paper at a comparatively low flow rate. Validity depends on a standard content of carbonaceous matter. Widely used in the UK in the past.

Standards and guidelines (WHO Air Quality Guideline values and EC Directive Limit Values can be found in Appendix 1.)

National Air Quality Standard:

50 μg/m³ PM_{10} as 24-hr running average. EPAQS also recommended that measures should be taken to aim to ensure that there is a decline in both peak and annual concentrations.

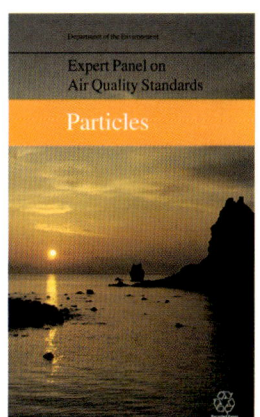

3.65 PM_{10} has only been measured in the UK since 1992 (levels of particles were previously measured as black smoke). Levels are generally between 10 and 45 µg/m³ with peak daily averages ranging from 70 to 160 µg/m³, although hourly means have exceeded 300 and 400 µg/m³ in certain cities. The annual mean is about 25-30 µg/m³ at urban monitoring sites.

Fig 3.9
Average smoke levels at Kew during October-March 1922-23 to 1970-71.

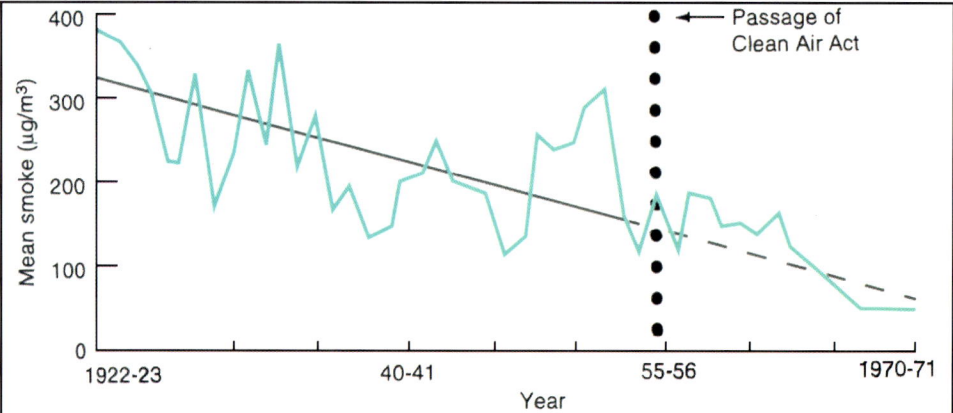

3.66   The EC limit value for black smoke (98%ile of daily means) was regularly exceeded at several UK monitoring sites during the 1970s (Belfast and some regions of Northern England), but the trend has been downwards due to wide-spread implementation of smoke control programmes and the limit value has rarely been exceeded during the 1980s and 1990s.

3.67   In comparison with the concentrations recorded when the domestic use of coal was the major source of particles, present-day concentrations in the UK are relatively very low.   Similarly low concentrations are also found in other West European cities.   In Eastern Europe much higher concentrations still occur as, to a lesser extent, they do in Southern European cities such as Athens.

Health effects of particles

3.68   Background.
A wide size range of particles occur in the air. These range from the molecular dimensions of 10 nm (0.01µm) to 100 µm. The finest particles are produced by the condensation of vapours and are called primary particles. These have a short life and aggregate to form particles of 0.1 - 1.0 µm diameter. Such secondary particles, which are also produced by the reaction of ammonia and sulphuric or nitric acid, are long-lived and may travel tens to thousands of kilometres. Rain is the main agent involved in removing these particles from the air. Larger particles, of 1-10 µm diameter, are mostly produced by physical means including wear and tear of road surfaces by tyres.

3.69   Particles of diameter greater than about 15 µm mostly do not enter the thoracic airways but are trapped in the nose and pharynx. Particles of less than about 4 µm diameter penetrate deeply into the lung. The efficiency of deposition of particles deep in the lung depends upon particle size. Recent deposition models have suggested that the deposition of very small (20 nm) particles is extraordinarily efficient with more than 60% of the inhaled mass of such particles being deposited in the alveolar region of the lung.

3.70 In considering particles it is essential to distinguish between mass and number of particles. Very fine (eg 20 nm diameter) particles weigh very little and in mass terms the deposited dose may be negligible. However, the number of particles deposited may be vast. A very large number of very fine particles may be contained in a small mass of material. Recent toxicological studies have demonstrated that low doses of ultrafine (less than 100 nm diameter) particles may have profound and unexpected effects in rats. The mechanism of these effects remains obscure and is currently under intensive study in a number of laboratories.

3.71 Until recently it was thought that the effects of particles on mortality, morbidity and the prevalence of respiratory disease seen during the smogs of 30 years ago showed a threshold at which no health effects could be observed. The 1987 WHO guidelines reflected this.

3.72 Though there is nothing new about pollution of the air by small particles, recent studies have demonstrated that what were considered safe levels of particles may have a significant effect on health. Studies demonstrating this relationship have accumulated since the late 1980s and most authorities now accept that the case for a relationship between mass concentrations of particles and effects on health is compelling. The findings are still remarkable. Much of this evidence comes from studies in North America, and the ability to detect relatively small effects depends on careful monitoring of ambient concentrations and indices of ill health and the use of advanced statistical analyses of complex data sets. Studies in the UK and other countries in Europe have tended to confirm the results of the US studies.

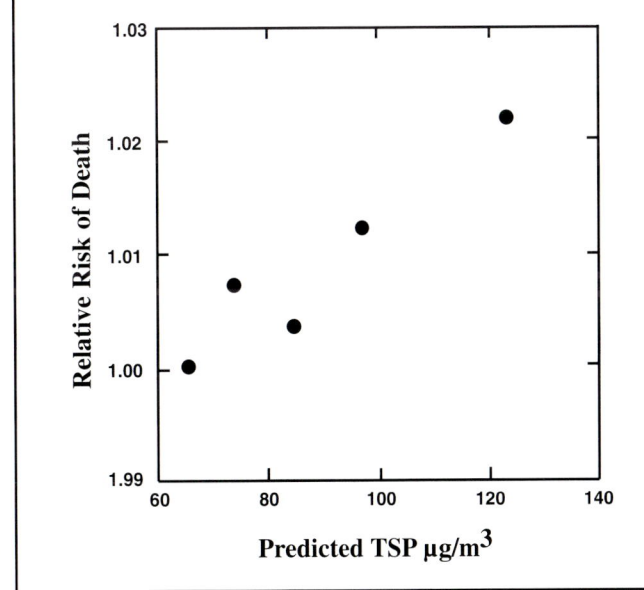

Fig 3.10
Relationship between predicted concentrations of particles and relative risk of death.

3.73 Two types of epidemiological study have driven the current thinking on the effects of particles on health. These are "time series studies" and "cohort studies". Each has produced important results though it has not been easy to correlate the results of the different types of study.

(a)   Time series studies

3.74   In these studies changes in indices of ill-health in a defined population, eg a city, have been related statistically to the day-to-day changes in concentrations of pollutants usually monitored at fixed sites in the city. Amongst the indices of ill-health monitored have been daily death rates, daily admissions to hospital, daily consumption of asthma relieving drugs (eg bronchodilator preparations), symptoms amongst those suffering from respiratory diseases such as asthma and standard indices of lung function. Marked day-to-day and season-to-season variations occur in these indices. Not all the reasons for these variations are understood though, for example, the effects of both unusually cold and unusually hot weather are well recognised. In addition, epidemics and mini-epidemics of infectious diseases exert a strong effect. For less obvious and sometimes administrative reasons, day of the week also has a significant effect on the daily hospital admission rates.

3.75   In studies of the effects of air pollutants on health these factors are described as confounding factors and the recent studies aim to take them into account when building a regression model relating changes in concentrations of pollutants to changes in indices of health. When these confounding factors have been taken into account a statistically significant association between day-to-day variations in mass concentrations of particles and indices of ill-health has been found to remain. Relationships with pollutants other than particles have also been discovered though, in general, these have tended not to be as strong as the relationship with particles or not to survive mathematical removal of the "particles term" from the complex multiple regression equation.

3.76   As in all studies of this kind the outcome is dependent upon the underlying assumptions of the statistical model deployed. In general, linear models without any assumption of a threshold of effect have been found to describe the data satisfactorily. In describing the results of such modelling exercises the slope (gradient) of the line which best describes the statistical relationship between "change in concentration of particles" and change in "index of ill-health" is usually quoted with the necessary and appropriate statistical confidence intervals. The confidence interval is often ignored in secondary accounts of the results of these studies.

3.77   As might have been anticipated not all studies have produced identical results, though the findings are strikingly consistent. A meta-analysis of the results of a number of studies was undertaken by Schwartz (one of the key workers in the area) and for daily mortality the following estimate of effect was produced:-

"A relative risk of death of 1.06 (95% CI = 1.05-1.07) was associated with a 100 μg/m^3 increase in concentration of total suspended particulate matter."

In terms of the commonly monitored PM_{10} this is equivalent to a 10% increase in daily deaths for each 100 μg/m^3 increase in PM_{10}. All concentrations are expressed as 24 hour average concentrations. It should be noted again that these studies shed no light on mechanisms of effect but simply report a consistent and statistically significant association. In addition, no information is available on the length of life lost. This may range from days to months and, perhaps, years.

3.78   If studies on mortality were the sole evidence linking particles and ill health there would perhaps be some difficulty in accepting the association as causal. However studies of other indices of ill-health have also yielded statistically significant associations and a number of studies have shown that effects on mortality and a range of indices of morbidity fit together in an impressive way. One meta-analysis suggested the following effects:

Estimates of effects of daily mean particulate pollution:

	% change in health indicator per each $10 \ \mu g/m^3$ increase in PM_{10}
Increase in daily mortality	
Total deaths	1.0
Respiratory deaths	3.4
Cardiovascular deaths	1.4
Increase in hospital usage (all respiratory	
admissions	0.8
Emergency department visits	1.0
Exacerbation of asthma	
Asthmatic attacks	3.0
Bronchodilator use	2.9
Hospital admissions	1.9
Increase in respiratory symptoms reports	
Lower respiratory	3.0
Upper respiratory	0.7
Cough	1.2
Decrease in lung function	
Forced expired volume	0.15
Peak expiratory flow	0.08

Table 3.2
Effects of PM_{10} on indicators of ill-health (from Dockery and Pope (1994) - a meta-analysis of studies on the effects of particles, each estimate is from a combination of two or more studies)

(b)   Cohort studies

3.79   For more than twenty years a study of health in six cities across the US has been under way. The study has related levels of ill health and the risk of death in the cities studied to long term average levels of pollutants. Confounding factors are of critical importance and adjustments for these are included in the analysis. The relationship of the adjusted death rates in each city corresponded to the level of PM_{10} measured in each city, with the ratio of adjusted death rate in Steubenville (the most polluted city) to that in Portage (the least polluted) being 1.26. Of the causes of mortality shown to be related to levels of particles, lung cancer is prominent along with cardio-respiratory diseases.

Fig 3.11
Relationship between Rate Ratio
for mortality and annual average
concentration of Fine Particles
(PM $_{2.5}$) in six US cities:

P	=	Portage
T	=	Topeka
W	=	Watertown
L	=	St Louis
H	=	Harriman
S	=	Steubenville

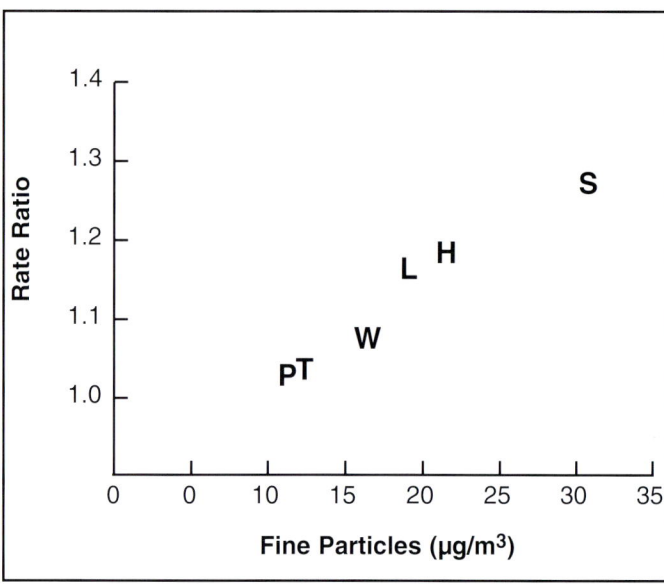

3.80 The classical "Six Cities Study" has recently been complemented by a study of more than 150 cities in the US and again more recently, by studies in Europe. The qualitative relationship between relative risk of death and long term average concentrations of particles discovered in the original Six Cities study has been confirmed in the expanded US study. Results from the European studies support, in general, the results of the US studies.

These studies add an additional dimension over that provided by the time series studies, as they suggest there may be long-term effects of exposure to particles. However, these studies have produced less consistent relationships in quantitative terms than the time series studies considered above.

Association and Causality

It is often fairly easy to show that some measure of ill-health, eg, the number of admissions to hospital per day, is associated with a possible cause such as the day-to-day variation in levels of air pollutants. To show real cause and effect, ie that a causal relationship exists, a number of guidelines or tests have been developed. These include looking at the consistency of the results of a number of different studies, the way in which the results of different studies fit together (coherence), whether there is a "dose-response relationship" such that as the proposed causal factor increases so does the effect and whether the sequence of events makes sense, ie the cause always precedes the effect.

Proof of causality is often impossible but by application of these and other criteria an expert judgement as to whether an association is likely to be causal can often be reached.

Interpretation of the results

3.81 Despite some residual concerns about confounding and the biological plausibility of the results, many experts are now agreed that it would be sensible to regard the demonstrated associations as causal and to make use of them in estimating the effects of particles on health. This task is not at all easy despite the consistency of the results. Difficulties in coming to clear conclusions include:

a. The problem of a possible threshold of effect has been raised and it has recently been suggested that epidemiological studies of the types described above are not suitable for detection of a threshold even if one were known to exist. This rather subtle point derives from the use of fixed site monitoring. It is known that individual exposure to particles will vary widely across a population. Thus though on a day when the fixed site monitor indicates a concentration of say 50 µg/m³ some individuals will have been exposed, at least for part of the 24 hours, to much higher concentrations and some to much lower concentrations.

b.   Epidemiological studies cannot demonstrate conclusively that the associations between levels of particles (or changes in levels of particles) and indices of health effects are, in fact, causal. A demonstration of causality will depend upon a clear understanding of the toxicological or other mechanisms of the effects observed. At present no generally accepted explanation has been put forward and further research is needed., though recent work has shown that free radicals may be involved in the toxicological effects of ultrafine particles.

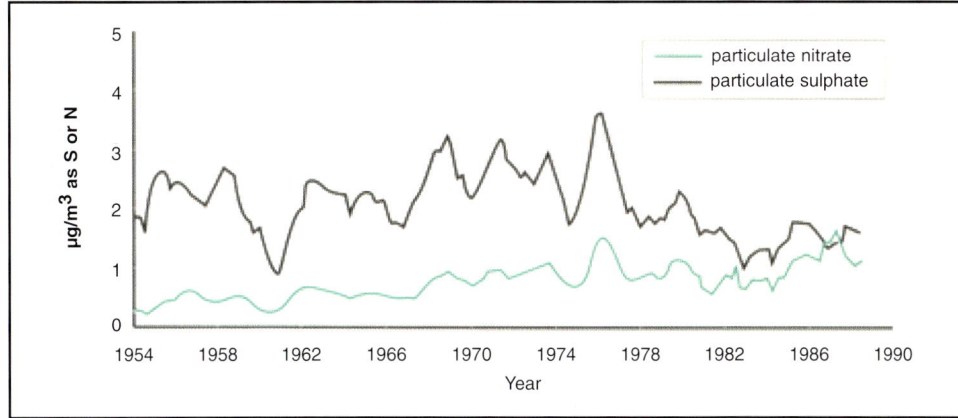

Fig 3.12
Three-Month running mean
Concentration of Particulate
Sulphate and Particulate Nitrate at
Chilton 1954 to 1988

c.   The analytical methods used in the time series studies inevitably generate a linear relationship. Some studies undertaken in areas where concentrations of particles are much greater than those commonly found in the US or in Western Europe have suggested a less steep relationship, ie a flatter line. Reanalysis of data collected in London in the 1960s has also demonstrated a less steep relationship than that summarised in the "100 µg/m³ increase in daily PM_{10} = 10% increase daily deaths" statement. It is now felt that the relationship over a wide range of concentrations may be curvi-linear and the relationship given above may not hold at concentrations greater than about 80 µg/m³.

d.   The described relationship does not shed any light on which component, or components, of the particle mixture may be most important in causing the effects described. This is of great importance in terms of planning an abatement strategy. The consistency of the results across different studies suggests that the precise chemical composition of the inhaled particles may not be critical to the effects reported, and this has led to the suggestion that the observed relationships between ill health and particle concentrations may be a consequence of the physical properties of the particles rather than their chemistry. This seems, at first glance, a major flaw in the work. However, not all effects need be dependent on chemical composition of the inhaled materials and this work suggests that there may be a "particle effect" in its own right.

e.   Examples which show that physical rather than chemical characteristics may be of importance are rare but one is provided by asbestos. Here, no examination of the chemical composition of asbestos fibres would lead one to say that such fibres were likely to be carcinogenic in small doses nor, certainly, to suggest that they might induce such a rare tumour as the pleural mesothelioma. It is now understood that the dimensions of the fibre, its low solubility and perhaps its surface properties are the important toxicological factors. Certainly, asbestos fibres have unusual properties, but it might be argued, so have particles less than 100 nm in diameter. That such particles might have toxicological properties as unpredictable as, though very different from, those of asbestos fibres now seems less implausible than it did a few years ago. This is an area of toxicology which is evolving rapidly and any views about possible mechanisms of effect must be regarded as tentative

f.   Important components of the particle mixtures are those relating to their acidity. Some workers have gone so far as to suggest that the various measures of particles available are all more or less good surrogates for airborne, probably particle transported, hydrogen ion. If this is true it will have profound implications for strategies designed to reduce the effects of particles on health.

3.82   Currently a precautionary approach has been taken in the UK, based on the prudent assumption of a causal link.

3.83   The DoE Expert Panel on Air Quality Standards (EPAQS) considered the evidence associating particle concentrations and indices of ill health concluding that it supported a causative link. It recommended a 24-hour standard for PM_{10} based on a concentration (50 $\mu g/m^{3)}$ at which, in its judgement, the risk of adverse health effects to any individual would be very small (of the order of one extra admission to hospital for treatment of lung disease per day in a population of a million). It also recommended action to reduce progressively the annual average concentrations of PM_{10}.

3.84   In response to the reports of COMEAP and EPAQS, the Government:

-     accepted the advice of COMEAP that it would be imprudent not to regard the associations between mortality and morbidity and particle concentrations as causal;

-     accepted EPAQS' recommendation that annual average levels of PM_{10} in the UK should continue progressively to be reduced;

-     adopted the value of 50 $\mu g/m^3$ running 24-hour mean as a specific objective to be achieved as the 99th percentile, that is on all but four days per year, by 2005.

3.85   The Government considered that these conclusions had five principal implications for policy:

-     reducing winter and summer peak levels of PM_{10} will be a priority;

- in winter time, when peaks are highest, vehicle emissions are crucial. The central elements of any strategy must therefore be technology-based measures to secure further long-term abatement of vehicle emissions, particularly from diesel vehicles;

- new vehicle emissions standards are relatively long-term in their impact. The Government intended therefore to take early action to reduce emissions from vehicles in urban centres;

- the most important particulate element in summer smogs is secondary aerosol. International action would be needed to tackle it successfully;

- given the Panel's conclusion that reduction of background levels will have a beneficial effect on health, and the possibility of local exceedences from specific sources, it would be right also progressively to continue to reduce emissions from all other relevant sources, including industry, quarrying, construction and domestic coal burning.

Volatile Organic Compounds (VOCs)

3.86   VOCs comprise a wide range of organic compounds. The major VOC is methane, with background concentrations of 1.6 ppm. Methane does not have significant toxicological properties and is not discussed here. Some VOC compounds are highly reactive with a short atmospheric lifespan, others can have a very long lifespan. The short lived compounds contribute substantially to atmospheric photochemical reactions and thus the formation of ozone.

Sources

3.87   Vehicle emissions are a major contributor, thus concentrations are generally higher in cities. Other major sources are solvent use (painting, dry-cleaning, degreasing etc) and industrial emissions.

Health effects

3.88   The major problem with VOCs is their potential for generating ozone (see above) but some VOCs such as benzene, 1,3-butadiene, polycyclic aromatic hydrocarbons (not all of which are in fact volatile), and dioxins, are hazardous in themselves.

Benzene

3.89   Benzene is a volatile organic compound (VOC).

Sources

3.90   There are no well-defined natural sources of benzene and most of the benzene observed at ground level in the northern hemisphere is likely to have resulted from human activities. About 80% comes from the benzene content and partial combustion of petrol in spark ignition cars. A further 5% comes from the storage and distribution of petrol and 1% from oil refining. Cigarette smoking is an important source for smokers and can be a significant source of benzene in the indoor environment.

3.91   New data on concentrations of benzene in the UK diet indicates that in non-smokers outdoor air is the major source of exposure.

Standards (WHO Guideline Values can be found in Appendix 1)

National Air Quality Standard:
5 ppb as a running annual average. (Target 1ppb)

UK levels

3.92   Ambient levels in the UK are between 1 and 40 ppb as hourly concentrations. Peak levels close to major emission sources (eg petrol stations) can be as high as several hundred ppb. Concentrations in outdoor air are generally at or below 5 ppb annual average, with urban annual average concentrations of approximately 1 - 3 ppb.

Health effects

3.93   Benzene is known to have carcinogenic and toxic effects. Studies of workers exposed to high concentrations of benzene (up to 100 ppm) for lengthy periods have demonstrated increased risks of leukaemia, principally of the acute myeloid type. Early signs of toxicity are anaemia, leucocytopenia or thrombo-cytopenia. Persistent exposure may cause damage to bone marrow leading to pancytopenia and exposure to levels in excess of 1000 ppm causes neurotoxic symptoms. No adverse effect on blood formation has been confirmed in humans following regular repeated exposure to benzene in air at concentrations below 25-30 ppm. EPAQS took 500 ppb (for a working lifetime) as a level at which one would not expect to demonstrate an increased level of leukaemia in feasible epidemiological studies. By the application of safety factors to account for whole life exposure and intersubject variability, a standard of 5 ppb, annual average concentration, was derived.

1,3-Butadiene

3.94   1,3-Butadiene is a volatile organic compound.

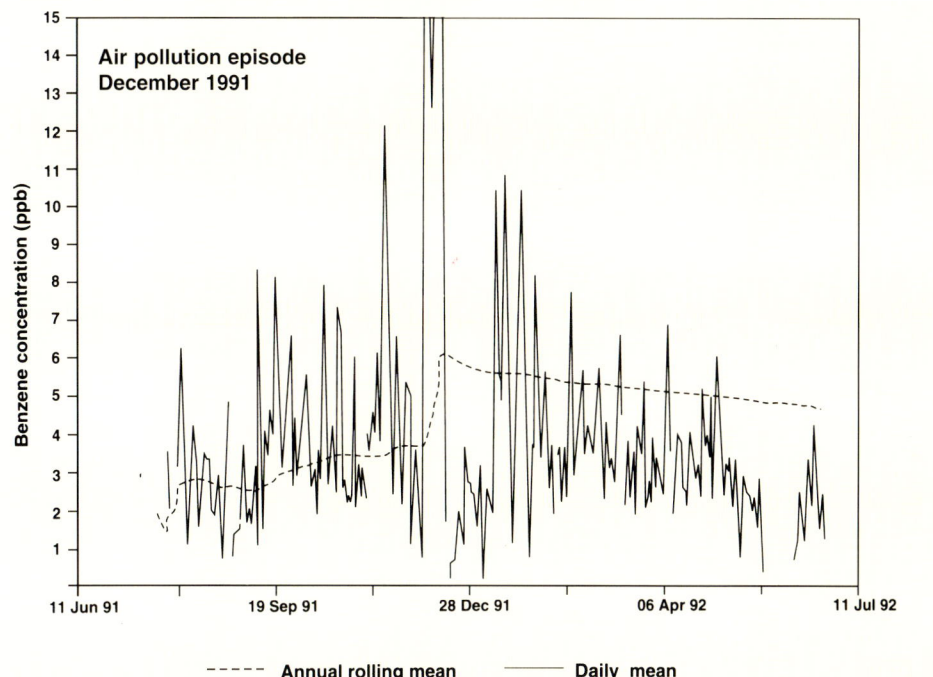

Air pollution episode December 1991

Y-axis: Benzene concentration (ppb)

X-axis: 11 Jun 91, 19 Sep 91, 28 Dec 91, 06 Apr 92, 11 Jul 92

- - - - - Annual rolling mean   ———— Daily mean

Fig 3.13
Rolling annual average and hourly average concentrations of benzene in London '91-'92.

Sources

3.95   Vehicle exhausts are the major UK source, but 1,3-butadiene is also produced by the combustion of fossil fuels and industrial processes. It is a widely used chemical intermediate in the manufacture of synthetic rubber tyres. Cigarette smoke is a major source in the indoor environment.

Standards

National Air Quality Standard:
1 ppb as a running annual average.

UK levels

3.96   Mean monthly concentrations in urban areas range from 0.2 to 1.0 ppb, with maximum hourly concentrations of up to approximately 10 ppb (Belfast). Peaks may also occur due to industrial emissions. Annual average concentrations are < 1ppb

Health effects

3.97   1,3-Butadiene is a potent genotoxic carcinogen in laboratory rodents. Epidemiological studies in humans occupationally exposed to 1,3-butadiene in rubber production plants suggest that exposure was associated with an increase in lymphomas and leukaemias. There were few data on exposure levels in these plants, but concentrations were in the range 1 - 10 ppm.

Polycyclic aromatic hydrocarbon (PAHs)

3.98   PAHs are a large group of organic compounds with two or more benzene rings. About 500 such compounds have been detected in air. The most studied member of the group is benzo(a)pyrene (BaP). PAHs are present in the atmosphere in gaseous form as well as adsorbed onto particulate matter (the bulk on respirable particles). The relative proportions of individual compounds in the mixture depends on the source of the PAHs.

Sources

3.99   PAH components are formed as a result of incomplete combustion. The burning of coal and coke production are still the major sources, but in urban areas vehicle emissions are likely to be the major source.

Standards and Guidelines

WHO Air Quality guidelines (1987): Owing to their carcinogenicity no safe level of PAHs can be recommended since all mixtures of PAHs contain BaP and other carcinogenic components, although not all PAHs have been shown to be carcinogenic.

WHO AQGS: 1995 Expert Group recommendations: Unit Risk (Risk associated with 1 ng/m^3 lifetime exposure) = 8.7 x 10^{-5} for BaP measured as a component of a PAH mixture.

UK Levels

3.100   BaP concentrations in most cities are in the range 1-5 ng/m^3 (levels were much higher when coal was used for domestic heating c. 100 ng/m^3). BaP concentrations of 40 ng/m^3 have been found in areas close to coke oven plants. Total PAH concentrations are generally in the range 10 - 150 ng/m^3, with occasional peaks of 300 - 700 ng/m^3. [These totals include all measurable PAHs, some of which are not thought to be carcinogenic].

Health effects

3.101   BaP is regarded as one of the most carcinogenic PAHs. There is evidence that, prolonged dermal exposure to PAHs (eg chimney sweeps, tar workers) leads to skin cancers. There is also an association between the occurrence of lung cancer and chronic exposure to PAHs in coke-oven workers exposed to a mixture containing approximately 30 µg/m^3 BaP, in coal gas workers and in employees in aluminium smelting plants.

3.102   Carcinogenicity is the only known toxic effect of environmental importance.

Carbon monoxide (CO)

3.103   CO is a colourless, odourless and tasteless gas. Once emitted into the atmosphere CO is slowly oxidised to CO_2.

Sources

3.104 Carbon monoxide is produced by the incomplete combustion of fossil fuel. Emissions are dominated by road transport, particularly petrol engines which are idling or decelerating, and this contribution has increased from 60% in 1970 to 90% in 1992 as traffic has increased. Badly flued heating and cooking appliances can be a significant source in the indoor environment and there are currently approximately 60 deaths per year and 500 sub-lethal incidents of poisoning in England and Wales by carbon monoxide coming from faulty or poorly maintained domestic gas appliances. Tobacco smoking is also a major source of carbon monoxide.

Standards and guidelines (CO)

National Air Quality Standard:
10 ppm as a running 8-hour average

Levels in UK

3.105 Levels in urban areas can be highly variable depending on weather conditions and traffic density. 8 hour averages in urban areas are generally < 10 ppm, but can be as high as 15 ppm. The relatively limited number of CO monitoring stations means that the UK distribution of CO is not accurately known, but there is a good correlation between CO and NO_x levels at urban sites. Peak levels in garage workshops can reach 500 ppm.

Health effects

3.106 When inhaled CO passes through the lungs and enters the blood, it binds to haemoglobin and disrupts the supply of oxygen to the tissues. Consequent reduced oxygen availability can lead to a wide range of health effects related to blood levels of carboxyhaemoglobin (COHb):

Table 3.3
Health effects related to blood levels of COHb

% COHb	Effect
2.3 - 4.3%	3-7% decrease in relation between work time and exhaustion in young healthy adults and ECG changes in patients with ischaemic heart disease
2.9 - 4.5%	decrease in exercise capacity in angina sufferers
5.0 - 5.5%	decrease in maximal oxygen consumption and exercise time in young healthy men
< 5.0%	no impairment of vigilance
5.0 - 7.6%	impaired vigilance
> 10%	headache, dizziness

In non-exposed populations, COHb levels are 0.4-0.7% resulting from endogenous formation of CO. Non-smoking populations exposed to environmental CO have COHb levels ranging from 0.5-1.5%. However, individuals in certain occupations, eg, policemen, traffic wardens, garage and tunnel workers, can have COHb levels up to 5%. Smoking increases COHb levels, heavy smokers achieving levels up to 10%.

3.107 Individuals at most risk from the effects of CO include those with cardiovascular or chronic respiratory problems, the elderly, pregnant women and young children.

Lead

3.108 Most airborne lead occurs as fine inorganic particles of submicron size (< 1 μm). Some 10% or less occurs as organic (ie alkyl) lead arising from petrol which has escaped combustion; the remainder is inorganic eg chlorides and carbonates.

Sources

3.109 In urban areas one of the major sources of lead in ambient air is emissions from petrol-engined road vehicles burning leaded fuel. However, in industrial areas, in the vicinity of processes such as secondary non-ferrous metal smelters, lead concentrations in ambient air can exceed those in urban areas.

Standards and guidelines

National Air Quality Standard:
0.5 μg/m^3 annual average
WHO Air Quality Guidelines (1987): Annual mean 0.5 -1.0 μg/m^3.

EC Directive: Limit value: 2 μg/m^3 annual mean.
Revision of WHO AQGS: 1995 Expert Group recommendations: 0.5 μg/m^3

UK levels

In rural areas lead levels are now < 0.05 μg/m^3, urban values are now <1 μg/m^3. Only one UK site has exceeded the EC limit value since 1985 (near an industrial works in Walsall) and since 1990 all sites have been in compliance with the EC limit. In 1993 the industrial area of Walsall was the only area in the UK to exceed the lower (0.5 μg/m^3) WHO guideline.

Health effects

3.110 Human exposure to lead is through inhalation of lead contaminated air and through the food chain, (which may contain lead from natural sources although deposition of airborne lead contributes to dietary intake). Blood lead concentrations are a good indicator of recent exposure from all sources and adverse health effects increase with increasing blood lead levels. The relative contribution of lead in ambient air to the total body burden in man is highly dependent on the intake from other sources, such as from food, water and the ingestion of dust.

3.111 The most sensitive body systems to the effects of lead are the haematopoietic system, the nervous system and the kidneys. Children are the most sensitive group and studies have indicated that children with high blood lead levels can suffer behavioural problems and lower IQ levels. There is no evidence of a threshold for these neurobehavioural effects. This is still slightly controversial, but generally accepted. Neurobehavioural effects have been noted at blood-lead levels of 10 μg/dl and there is still a proportion of children in the UK at or close to this level. Recent data suggest that about 5% of children in the UK have blood lead concentrations of >10 μg/dl.

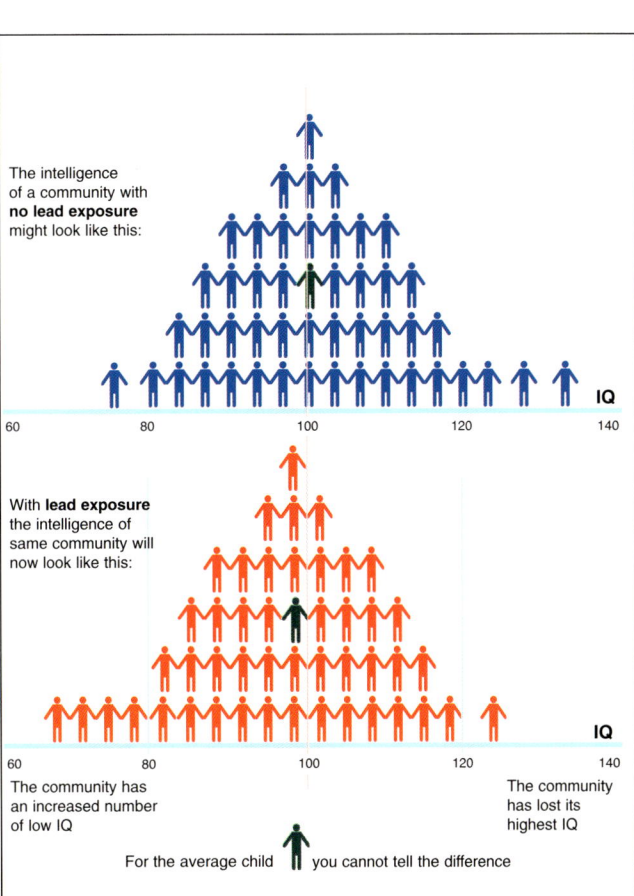

The intelligence of a community with **no lead exposure** might look like this:

With **lead exposure** the intelligence of same community will now look like this:

The community has an increased number of low IQ

The community has lost its highest IQ

For the average child you cannot tell the difference

Fig 3.14
Distribution of IQ in a non lead exposed and in a lead exposed population

Cadmium

3.112 Cadmium is mainly present in the atmosphere in particulate form. Human exposure to cadmium is via inhalation and through the food chain.

Sources

3.113 Major sources in the UK are oil and coal combustion, metal production, and battery manufacture. Tobacco smoking can make a significant contribution in the indoor environment.

Standards and guidelines

WHO Air Quality Guidelines (1987): Limited evidence of human carcinogenicity, no risk estimate based on carcinogenic effects. Guideline values based on non-carcinogenic effects as follows: rural areas, present levels of 1-5 ng/m³ should not be allowed to increase. Urban areas (no agriculture and industrialised): levels of 10 - 20 ng/m³ tolerable.

Present levels should not be allowed to increase.

UK levels

3.114 Yearly mean concentrations: rural: < 1-5 ng/m^3, urban: 5-15 ng/m^3, industrial: 15-50 ng/m^3. Short term values near metal processing plants can be as high as 5-11 µg/m^3.

Health effects

3.115 Airborne cadmium is absorbed in the lungs and is concentrated in the liver and kidneys, where it displaces zinc from a number of zinc-containing metallo-enzymes. The cadmium binds irreversibly with the active sites, destroying normal functioning.

3.116 In terms of long term low level exposure, the kidney is the critical target organ, with damage to renal function as cadmium accumulates above about 200 mg/kg in the kidney. Calculations suggest that this level could accumulate after continuous exposure to about 3 µg/m^3 in air for 50 years. [Guideline levels (see above) are lower than this to take account of deposition of cadmium from air leading to exposure to cadmium via food.]

3.117 Acute and chronic respiratory effects may be expected to result from exposure to 1 mg/m^3 and 20 µg/m^3 (for 20 years), respectively.

Environmental tobacco smoke

3.118 Environmental tobacco smoke (ETS) is a mixture of thousands of compounds in vapour and particulate phases. ETS cannot be measured as a whole. Measurements are normally carried out via marker compounds such as nicotine and respirable suspended particulates (RSPs).

Indoor levels of ETS

3.119 Nicotine levels in homes with smokers range from 1-10 mg/m^3. Levels are higher in restaurants and bars and confined spaces such as inside cars. Levels of ETS associated RSPs of up to 500 µg/m^3 have been reported in homes with smokers. Levels in restaurants and bars can exceed 1 mg/m^3.

Health effects

3.120 The pattern of health effects in adult non-smokers from exposure to ETS is consistent with the effects known to be associated with active smoking. Chronic exposure to ETS is associated with increased mortality due to lung cancer, and increased risk of mortality and morbidity from cardiovascular disease. The World Health Organisation has estimated that 9-13% of all cancer cases can be attributed to ETS in a non-smoking population of which 50% is exposed to ETS. In 1993, the Department of Health's Committee on Carcinogenicity of Chemicals in Food, Consumer Products and the Environment considered ETS and came to the following conclusion:

> "The consensus from the the published literature was that there was a small, statistically significant risk of lung cancer among non-smokers exposed over a substantial part of their lifetime to

environmental tobacco smoke. The Committee agreed with the conclusion of the Independent Scientific Committee on Smoking and Health that the relative risk among such people was in the range 1.1 to 1.3, accounting for several hundred lung cancer deaths per annum in the UK".

3.121   In infants and young children, exposure to ETS has been reported to increase the risk of asthma, pneumonia, bronchitis, bronchiolitis and fluid in the middle ear. ETS exposure also increases the severity and frequency of asthma attacks. In the case of asthma, smoking in pregnancy retards fetal lung development while exposure after birth increases the frequency and severity of asthma attacks.

Formaldehyde

Sources

3.122   Formaldehyde levels indoors are determined by emissions from many products including insulating materials (such as urea-formaldehyde foam), particle board and other construction materials, furniture made from chipboard or plywood, water based paints, fabrics, household cleaning agents and disinfectants. This reactive pollutant is also present in combustion products and cigarette smoke.

Indoor levels of formaldehyde

3.123   Studies have shown that airborne concentrations of formaldehyde in homes depend on the age of the source material and ventilation, temperature and humidity. Active and passive cigarette smoking also contribute to total personal exposure. A study of 174 homes in Avon has shown the annual mean indoor level of formaldehyde to range from 0.020 to 0.025 mg/m^3 according to the room sampled. Significant differences were demonstrated between individual homes, with the highest mean concentration (for 12 monthly readings) being 0.076 mg/m^3 and the lowest 0.007 mg/m^3. This is higher than the outdoor mean concentration of 0.002 mg/m^3. A comprehensive European survey in homes and schools reported levels between 0.01 and 0.1 mg/m^3. Overall levels in UK homes do not differ from those in dwellings elsewhere in the world.

3.124   Increased formaldehyde levels have been demonstrated in newer homes, homes with integral garages, homes with new furnishings and recently decorated homes. The mean formaldehyde concentration increases the newer the dwelling and the mean concentration in homes built since 1982 is about three times higher than those built before 1919. This has been suggested to be due to off-gassing from new materials, increases in the number of formaldehyde sources in the home, or increased 'tightness' of homes as a consequence of energy conservation measures. Seasonal fluctuations may be due to changes in ventilation or to use of different household products at different times of year.

Health effects

3.125 Whilst high concentrations of formaldehyde can serve as a small molecular weight sensitizing chemical, eg in radiographers or pathologists, the sensory, irritant and respiratory effects of formaldehyde are more relevant at the low concentrations found in modern homes. An assessment of the health effects has concluded that chamber studies with normal adults or those with pre-existing asthma have not demonstrated any dysfunction at mean formaldehyde levels currently found in homes or even at levels several times higher. Epidemiological studies have not demonstrated any increases in respiratory symptoms or decreases in lung function at estimated current levels in homes. There is no evidence to suggest that the current domestic exposures to formaldehyde pose a carcinogenic risk. However, for the protection of health, exposure to formaldehyde in the domestic environment should remain at or below current levels.

Standards and Guidelines

WHO Air Quality Guidelines: revised 1996, 30 minute average concentration: 0.1 mg/m^3.

This guideline is intended to prevent sensory irritation in the general population.

SECTION 4

RESEARCH

- **Introduction**
- **DH policy research programme**
- **DH/DoE/MRC programme of research**

RESEARCH

INTRODUCTION

4.1 The Government is taking an active part in a wide range of relevant R&D activity in this and other fields. The Department of Health's own research is intended to help clarify factors affecting public health and NHS research is expected to identify those priorities which have the most direct impact on the health service. The Department of Health is involved in research in a number of ways:

- On its own behalf, to provide information from research to assist policy development directed at factors which affect public health.

- Through the NHS R&D Strategy, which is intended to ensure that the service develops the research competence to review critically its own needs, contributing to the nation's health through improved quality, effectiveness and cost-effectiveness of the NHS.

- Through links with other Government and non-Government research funders to ensure a fruitful exchange of research problems and results among the various bodies funding health research.

- Through its membership of the Respiratory Research Liaison Committee. This Committee has provided a basis for development of inter-sectoral work with a respiratory disease focus, through representation from the major organisations with a research interest in the field including, Government research, charities and industry. The future activities of that Committee are currently being reviewed in the context of the developing plans for the National Forum.

4.2 Research into the health effects of air pollutants is commissioned by a number of bodies, including the Medical Research Council, Government departments and charities and other organisations.

IEH reports on air pollution

DH POLICY RESEARCH PROGRAMME

4.3 The Policy Research Programme is the main vehicle by which the Department of Health funds research which supports the development and implementation of policy within the Department. It is separate from the NHS research programmes, which focus on work in support of the delivery of services by the NHS.

Research underway

University College London Medical School

Impact of air pollution and temperature on GP consultations. Effects of ozone, climatic conditions on daily GP consultation rates; levels of oxides of nitrogen and sulphur, particulates and pollen. Different respiratory conditions in different age groups, including asthma. South east, Wessex and South western regions.

Centre for Environmental Technology, Imperial College

Work to assess personal exposure to air pollutants and severity of hayfever symptoms in London traffic wardens - association between exposures and severity of nasal, eye and respiratory symptoms.

A mortality study of cadmium exposed workers with particular reference to lung and prostatic cancer (Imperial College)

OTHER RESEARCH

Several projects which examine census-linked data include reference to asthma prevalence, including analysis of data of 1958 birth cohort (London University Institute of Child Health)

Characterisation of fine particulate matter (PM_{10}) in airborne samples and human lung tissue (University of Wales)

DH/DoE/MRC PROGRAMME OF RESEARCH PROPOSALS ON AIR POLLUTION AND RESPIRATORY DISEASE

4.4   Air pollution research is a priority within the DH centrally commissioned research programme and together with DoE and the MRC the Department issued a call for research proposals on air pollution and respiratory disease in October 1994. This was co-ordinated by the MRC Institute for Environment and Health (IEH) at Leicester and covers ten key areas, which had been previously identified by discussion with the Departments and their expert advisory groups, such as COMEAP. The programme represents a possible £5M of research grants.

4.5   The key areas are:

Concerning immediate (acute) effects of air pollution on health and remote (acute or chronic) effects, three areas for each:

- Does this health effect occur under the air pollution conditions currently present in the UK and how large is the effect?

- Which pollutants and modifying factors may be responsible?

- Which physiological or pathological mechanisms could be responsible for the effects reported?

The other four areas are concerned with the development of personal monitoring methods and the relationship of personal exposure to fixed site monitoring and in defining the nature of the dose-response curve for a range of pollutants.

4.6   This approach is intended to coordinate research priorities in this field, minimise the possibility of overlap, duplication or gaps in the research, while allowing DH, DoE and MRC to select and fund specific, appropriate research projects. Over 200 proposals were received and these were subjected to an initial sift in February 1995. Proposals which were successful at this stage have now been handed to the appropriate funding body to be followed up in that body's usual manner by the Departments of Health and of the Environment.

4.7   A list of research projects, including title, workers and abstracts, funded under the initiative is attached as Appendix 2 of this handbook.

SECTION 5

INTERNATIONAL ACTIVITIES

- **EU Air Quality Directives**
- **WHO Air Quality Guidelines**

EU AIR QUALITY DIRECTIVES

5.1   A number of EC Directives concern Air Quality in the UK. These and some National regulations are listed at the end of this section, taken from the first report of the DOE Quality of Urban Air Review Group. It will be seen that the first Directives concerned emissions from motor vehicles. Later Directives dealt specifically with individual pollutants and established Limit Values and, in some cases, Guide Values for these pollutants. Details of individual Limit and Guide Values are provided in Appendix 1.

EU AMBIENT AIR QUALITY ASSESSMENT AND MANAGEMENT DIRECTIVE

5.2   The Ambient Air Quality Assessment in Management Directive reached common position last year. It received its second European Parliament reading in May 1996 and was formally adopted in September 1996 and published in the Official Journal in November. The Directive requires transposition of its provisions within eighteen months of publication. Proposals for "daughter Directives" for the first four pollutants are expected by mid 1997.

5.3   The Directive sets out the framework for air quality policy in Europe, giving the general monitoring strategy and the concerns to be addressed when setting limit values. It lists thirteen pollutants to be covered: sulphur dioxide, nitrogen dioxide, particles (given as two in the directive - fine particles including PM_{10} and "Suspended Particulate Matter"), lead, ozone, benzene, carbon monoxide, polyaromatic hydrocarbons, cadmium, arsenic, nickel and mercury.

5.4   Each of these pollutants is to be covered by a "daughter" directive setting out the actual Limit Value, specific monitoring requirements and other issues particular to the pollutant. The Directive commits the Commission to producing daughter directives for the first four by 31 December 1996, ozone by March 1998 (see below) and the rest by 31 March 1999. Work on the first four daughter directives is well under way. Technical Working Groups have prepared position papers on each of these pollutants. Each Group had members from four or five Member States, the Commission (DGXI and DGXII), the European Environment Agency, the World Health Organisation, the Joint Research Centre, European NGOs and industry. Their reports will pass to the Commission via a Steering Committee where all Member States, plus the other aforementioned interested parties, are represented. The Commission will use these reports as the basis of their proposals for daughter directives. The timetable for publication of the proposals has, however, been pushed back to mid-1997.

5.5   Technical working groups now exist for ozone and benzene. Proposals for these pollutants and carbon monoxide are due at the end of 1997. The exact timetable for the remaining pollutants has yet to be decided. The Directive also has scope for other pollutants to be added at a later date.

World Health Organization: Air Quality Guidelines for Europe

5.6   In 1987 WHO published the first edition of the Air Quality Guidelines for Europe. These dealt with some 28 compounds and provided guidance on levels at which, for the great majority of people, no adverse effects on health would be expected to occur. A process of revising these guidelines was begun in 1993 and compounds have been reviewed in groups. A number of substances not included in the original edition have been added during this revision.

REGULATIONS RELEVANT TO UK URBAN AIR QUALITY, 1956-1992

1956 **Clean Air Act** - Introduced Smoke Control Areas and grants for conversion to smokeless fuels. Controlled chimney heights. Prohibited emissions of dark smoke from chimneys with some exceptions.

1968 **Clean Air Act** - Extended the Smoke Control provisions. Extended the prohibition of dark smoke emissions.

1970 **EC Directive 70/220/EEC** - Relating to measures to be taken against air pollution by gases from positive ignition engines of motor vehicles. Limited emissions of carbon monoxide and hydrocarbons from petrol cars. Introduced into European legislation the requirements of UN Economic Commission for Europe (ECE) Regulation 15. Came into force in 1971.

1971 **The Performance of Diesel Engines for Road Vehicles BSI AU 141a** - Controlled black smoke from heavy duty vehicles.

1972 **EC Directive 72/306/EEC** - Measures to be taken against emissions of pollutants from diesel engines for use in motor vehicles. Limited black smoke emissions from heavy duty vehicles. Essentially extended UK BSI AU 141 requirements throughout the EC.

1974 **Control of Pollution Act** - Allowed for the regulation of the composition of motor fuels; and limited sulphur in fuel oil.

1974 **EC Directive 74/209/EEC** - Amending 1970 directive relating to measures to be taken against air pollution by gases from positive ignition engines of motor vehicles. Reduced limit values for carbon monoxide and hydrocarbons emissions from petrol cars in line with UN ECE Regulation 15.01. Came into force in 1975.

1975 **EC Directive 75/441/EEC** - Established a common procedure for the exchange of information between the surveillance and monitoring networks based on data relating to sulphur dioxide and smoke. Set up a procedure for exchanging air quality information between Member States.

1975 **EC Directive 75/716/EEC** - Relating to the approximation of the laws of Member States on the sulphur content of certain liquid fuels. Defined two types of gas oil (diesel and heating oil). Introduced in two stages sulphur limits for these fuels.

1977 **EC Directive 77/102/EEC** - Amended 1970 directive relating to measures to be taken against air pollution by gases from positive ignition engines of motor vehicles. Reduced limit values for carbon monoxide and hydrocarbons. Introduced limit values for nitrogen oxides emissions from petrol cars in line with UN ECE Regulation 15.02. Came into force in 1977.

1978 **EC Directive 78/665/EEC** - Amended 1970 directive relating to measures to be taken against air pollution by gases from positive ignition engines of motor vehicles. Reduced limit values for carbon monoxide, hydrocarbons, and nitrogen oxides emissions from petrol cars in line with UN ECE Regulation 15.03. Came into force in 1979.

1978 **EC Directive 78/611/EEC** - Concerning the lead content of petrol. Limited the maximum permissible lead content of petrol at 0.4 g l^{-1}. Came into force in 1981.

1980 **EC Directive 80/779/EEC** - Air quality limit values and guide values for sulphur dioxide and suspended particulates.

1981 **The Motor Fuel (Lead Content of Petrol) Regulations** - Limited the maximum amount of lead in petrol to 0.4 g l^{-1}.

1982 **EC Directive 82/459/EEC** - Repealed Directive 75/441/EEC and established a reciprocal exchange of information and data from networks and individual stations measuring air pollution within Member States. Set up a procedure for exchanging air quality information.

1982 **EC Directive 82/884/EEC** - Limit value for lead in the air.

1983 **EC Directive 83/351/EEC** - Amended 1970 directive relating to measures to be taken against air pollution by gases from engines of motor vehicles. Reduced limit values for carbon monoxide, hydrocarbons and nitrogen oxides emissions from petrol cars in line with UN ECE Regulation 15.04. Introduced limits for diesel engines for first time. Due to come into force in 1984 but was never introduced into British legislation.

1984 **EC Directive 84/360/EEC** - A framework directive on combating air pollution from industrial plants. The introduction of a common framework for reducing air pollution from industrial plants in the Community. Came into force in 1987.

1985   **EC Directive 85/203/EEC** - Air quality standards for nitrogen dioxide.

1985   **EC Directive 85/210/EEC** - The approximation of Member State legislation on lead content of petrol, and the introduction of lead-free petrol. Allowed for the introduction of unleaded petrol. Limited the benzene concentration of petrol to 5% by volume.

1985   **The Motor Fuel (Lead Content of Petrol) Regulations** - Limited the maximum amount of lead in petrol to 0.15 g l^{-1}. Came into force in December 1985.

1987   **EC Directive 87/217/EEC** - Prevention and reduction of environmental pollution by asbestos. Controlled the pollution of air, water and land by asbestos from all significant point sources.

1987   **EC Directive 87/219/EEC** - Amended Directive 75/716/EEC. Limited sulphur content of all gas oil to 0.3% by weight.

1987   **EC Directive 88/76/EEC** - Amended 1970 directive relating to measures to be taken against air pollution by gases from engines of motor vehicles. Sets stringent emission standards for carbon monoxide, hydrocarbon and nitrogen oxides together and nitrogen oxides from large cars (> 2 litres), with less stringent requirements from medium and small cars.

1987   **EC Directive 88/77/EEC** - The approximation of the laws of the Member States relating to the measures to be taken against the emission of gaseous pollutants from diesel engines for use in vehicles. Controlled emissions of gaseous pollutants from heavy duty vehicles.

1988   **EC Directive 88/609/EEC** - Limited emissions of certain pollutants into the air from large combustion plants. Limited emissions of sulphur dioxide, nitrogen oxides and particulates from power stations and other large combustion plants.

1988   **EC Directive 88/436/EEC** - Amendment of 1970 directive relating to measures to be taken against air pollution by gases from engines of motor vehicles. Controlled particulate emissions from diesel cars.

1989   **EC Directive 89/369/EEC** - Directive on air pollution from new municipal waste incinerators. Set emission limits on new waste incinerators.

1989   **EC Directive 89/429/EEC** - Directive on air pollution from existing municipal waste incinerators. Set limits on emissions from existing waste incinerators.

1989   **EC Directive 89/427/EEC** - Limit values and guide values of air quality for sulphur dioxide and suspended particulates. Harmonised measurement methods.

1989   **EC Directive 89/458/EEC** - Amendment of 1970 directive relating to measures to be taken against air pollution by gases from engines of motor vehicles. Introduced limits for emissions of carbon monoxide and hydrocarbons and nitrogen oxides from small cars (< 1.4 litres). Mandatory for the first time.

1989   **The Air Quality Standards Regulations 1989** - Statutory Instrument No 317. Formally brought into UK legislation the limit and guide values for sulphur dioxide and suspended particulates, lead in air and nitrogen dioxide set by EC Directives.

1990   **The Motor Fuel (Sulphur Content of Gas Oil) (Amendment) Regulations** - Brought into UK legislation the requirements of EC Directive 87/219/EEC that limits the sulphur content of gas oil to 0.3% by weight.

1990   **The Oil Fuel (Sulphur Content of Gas) Regulations** - Brought into UK legislation the requirements of EC Directive 87/219/EEC that limits the sulphur content of gas oil to 0.3% by weight.

1990   **Environmental Protection Act** - Brings many smaller emission sources under air pollution control by local authorities for the first time.

1991   **EC Directive 91/441/EEC** ('The Consolidated Directive') - Amendment of 1970 directive relating to measures to be taken against air pollution by gases from engines of motor vehicles. Introduced mandatory emission requirements that effectively require the fitting of closed loop three way catalysts to all new petrol cars from 1993 and new limit values for diesel cars.

1991   **EC Directive 91/542/EEC** - Amendment of Directive 88/77/EEC on the approximation of the laws of the Member States relating to the measures to be taken against the emission of gaseous pollutants from diesel engines for use in vehicles. Introduced a two stage reduction in emission from heavy duty diesel vehicles (to come into effect in 1992 and 1995). Introduced emission controls for particulates for the first time.

1991   **The Road Vehicles (Construction and Use) (Amendment) Regulations 1991** - Set standards for in-service emissions of carbon monoxide and hydrocarbons to be included in the MOT test for petrol cars and light goods vehicles.

1992   **Agreed EC Directive** but unpublished at the time of writing this report. Amends Directive 87/219/EEC. Limits sulphur content of diesel to 0.05% by weight.

1992   **EC Directive 92/72/EEC** - Air Pollution by Ozone. Establishes a harmonised procedure for monitoring, exchange of information and informing and warning the public on ozone pollution.

1996   **EC Directive 96/62/EEC** - Ambient air quality assessment and management ("Framework Directive").

APPENDICES

Appendix 1

International Standards and Guidelines

Sulphur Dioxide

Table A1.1

WHO Air Quality Guidelines (1987)

10 min: 175 ppb (501 µg/m^3)
1 hr: 122 ppb (349 µg/m^3)

Table A1.2

Revision of **WHO AQGS: 1996 Expert Group recommendations**
(1 hr guideline and link with smoke abandoned):

10 min: 175 ppb (500.5 µg/m^3)
24 hr: 45 ppb (128.7 µg/m^3)
Annual: 17 ppb (47 µg/m^3)

Table A1.3

EC directive: Limit values:
Annual median of daily means:

30 ppb (UK equiv 34 µg/m^3) if smoke > 40 µg/m^3
45 ppb if smoke < 40 µg/m^3

Winter median of daily means:

48.8 ppb (UK equiv 51 µg/m^3) if smoke > 60 µg/m^3
67.5 ppb if smoke < 60 µg/m^3

98%ile of daily means:

93.8 ppb (UK equiv 1281 µg/m^3) if smoke > 150 µg/m^3
131.3 ppb if smoke < 150 µg/m^3

Guide values:

Annual average: 15 - 22.9 ppb (43-65 µg/m^3)
Daily average: 37.5 - 56.4 ppb (107 - 161 µg/m^3)

Nitrogen Dioxide

Table A1.4

WHO Air Quality Guidelines (1987)

1 hr: -   210 ppb (385 µg/m^3)

24 hr: -   80 ppb (150 µg/m^3)

Table A1.5

Revision of **WHO AQGS: 1996 Expert Group recommendations**

(24 hour guideline abandoned):

1 hr: 110 ppb (207 µg/m^3)

Annual: 40 µg/m^3 (21 ppb)

The change in the 1 hour guideline reflects the increasing concern that NO$_2$ may play some adjuvant role in asthma both in terms of the provocation of attacks by allergens and, though less likely, in terms of the initiation of the disease. The introduction of an annual guideline reflects the results of epidemiological studies which show a negative association between long term average concentrations of NO$_2$ and indices of lung function. There is some room for doubt in the interpretation of the results of these studies as regards which pollutant or combination of pollutants is responsible for the described effect and the annual guideline is less firmly founded than some other WHO Air Quality Guidelines.

Table A1.6

EC Directive Limit Values

Limit value:	98%ile hourly mean:	104.6 ppb (197 µg/m^3)
Guide values:	98%ile hourly mean:	70.6 ppb (133 µg/m^3)
	50%ile hourly mean:	26.2 ppb (49 µg/m^3)

Ozone

Table A1.7

WHO Air Quality Guidelines (1987)

1 hr: 76 - 100 ppb (152 - 200 µg/m^3)

8 hr: 50 - 60 ppb (100 - 120 µg/m^3)

Table A1.8

Revision of **WHO AQS: 1996 Expert Group Recommendations**:

8 hr: 60 ppb (120 $\mu g/m^3$) (1 hr guideline abandoned) + tables of exposure-response for 1 hr and 8 hr, because there was little evidence of a threshold for effects

Table A1.9

EC directive:

Health protection threshold:	8 hr:	55 ppb (110 $\mu g/m^3$)
Population information threshold	1 hr:	90 ppb (180 $\mu g/m^3$)
Population warning value:	1 hr:	180 ppb (360 $\mu g/m^3$)

Particulate Matter

Table A1.10

WHO Air Quality Guidelines (1987)

Based on epidemiology and therefore done in conjunction with SO_2.

Short term (24 hr): SO_2 125 $\mu g/m^3$ (44 ppb); BS 125 $\mu g/m^3$ (TSP 120 $\mu g/m^3$; TP 70 $\mu g/m^3$)

Long term (1 yr): SO_2 50 $\mu g/m^3$ (17.5 ppb); BS 50 $\mu g/m^3$) (No figures suggested for TSP/TP)

Recommendations made to WHO by expert group 1994 (link with SO_2 abandoned): Non-threshold effect. Provides a dose-response table for PM_{10}, $PM_{2.5}$, and aerosol sulphate, for mortality and other outcomes, eg:

Health effect indicator	Estimated change in daily average concentration needed for given effect ($\mu g/m^3$)		
	Sulphates	**$PM_{2.5}$**	**PM_{10}**
Daily mortality:			
5% change	8	29	50
10% change	16	55	100
20% change	30	110	200
Hospital admissions - respiratory conditions:			
5% change	8	10	25
10% change	16	20	50

Table A1.11

EC directive: (EC Black Smoke not identical to UK, UK BS = 0.85 EC BS)

Limit values:
Annual median of daily means: 80 µg/m³ (UK BS equivalent 68 µg/m³)

Winter median of daily means: 130 µg/m³ (UK BS equivalent 111 µg/m³)
98%ile of daily means throughout year: 250 µg/m³ (UK BS equivalent 213 µg/m³)

Guide values:
Annual average:40 - 60 µg/m³ (UK BS equivalent 34 - 51 µg/m³)
Daily average:100 - 150 µg/m³ (UK BS equivalent 85 - 128 µg/m³)

US primary standard for protection of human health:
24 hr mean not to be exceeded more than once a year: 150 µg/m³ PM_{10}
Annual arithmetic mean: 50 µg/m³ PM_{10}

Benzene

WHO Air Quality Guidelines (1987)

No safe level for airborne benzene can be recommended, as benzene is carcinogenic to humans and there is no known safe threshold level. WHO calculated that at an air concentration of 1 µg/m³ (equivalent to 0.313 ppb) the estimated lifetime risk of leukaemia is 4×10^{-6}. There are many problems inherent in this mathematical estimate and the EPAQS standard is set pragmatically at a level at which health risks are exceedingly small and effectively merges with the background level of risk among non-exposed groups..

Revision of WHO AQGS: 1995 Expert Group recommendations

Unit Risk (risk associated with 1 µg/m³ lifetime exposure) = 4.4×10^{-6} - 7.5×10^{-6}.

Carbon Monoxide

Table A1.13

WHO Air Quality Guidelines (1987)

15 min:	87 ppm (109 mg/m³)
30 min:	50 ppm (62.5 mg/m³)
1 hr:	25 ppm (31.25 mg/m³)
8 hr:	10 ppm (12.5 mg/m³)

The WHO guidelines are calculated such that a normal subject engaging in relatively heavy work will not exceed a COHb level of 2.5%.

Revision of **WHO AQGS: 1996 Expert Group recommendations** - no change
EC directive: None as yet

Appendix 2

Research Projects

Author(s):	Dr Suzanne Moffatt, Lecturer in Social Epidemiology
	Dr Christine Dunn, Lecturer in Geography

Project Title: Public Awareness of Air Quality and Respiratory Health Assessing the Impact of Health Advice

Official Address: Department of Epidemiology and Public Health
The Medical School
University of Newcastle upon Tyne
Newcastle upon Tyne

Project Abstract: This research aims to study, in depth, public awareness and perceptions of the relationship between ill-health and air quality. The intention is to examine the effect of local and national air quality information and subsequent health advice and relate this to policy needs. The study will examine how individuals and communities drawn from   different socio-economic groups prioritise air pollution and health issues. Furthermore, it will investigate the impact of information about air quality and public health on behaviour.

Working class and middle class populations resident in north east England will be studied; those in one town living at varying distances from large-scale steel and petrochemical industries and comparable groups in another town without nearby industry.

The sample size will comprise 5,002 adults aged 17-80, stratified by social class and location.

A postal questionnaire will be sent to the entire sample, and in-depth interviews with a sub-sample of 100 respondents with respiratory illness and 100 without are to be undertaken.

Author(s): Dr Anthony James Frew, Senior Lecturer in Medicine, University of Southampton
Dr Frank James Kelly, Senior Lecturer in Biochemistry, St Thomas's Hospital
Dermot Crean, Surgeon Captain, Institute of Naval Medicine

Project Title: Feasibility Study of Adapting the EMU of the INM for Studies of Human Exposure to Pollutant Gases

Official Address: Department of Medicine, Centre Block
Southampton General Hospital
Tremona Road,
Southampton, SO16 6YD

Project Abstract: The Environmental Monitoring Unit (EMU) is a self-contained research laboratory which can house up to 12 subjects for several days at a time. The authors propose to adapt the EMU for medium term exposure to ozone and nitrogen dioxide. Time will be purchased on the EMU, and a suitable ozone generating system installed. A series of trials will then be conducted to evaluate the stability of ozone concentrations in the range 100-400 ppb, together with sampling of different sites within the EMU to confirm even distribution of gas within the chamber. Once completed, the modified EMU will be made available for clinical trials. These will include studies of the physiological effects of ozone exposure, as well as the cellular and biochemical consequences of pollutant exposure. Separate application will be made for these later trials.

Author(s):	Professor H R Anderson
Project Title:	Effects of Air Pollution on Daily Mortality, Admissions and General Practitioner Consultations in London
Official Address:	St George's Hospital Medical School Department of Public Health Sciences Cranmer Terrace London SW17 0RE

Project Abstract:

The aims are to investigate whether air pollution has short-term effects on health in Greater London, and, if so, to quantify the nature and scale of any such effects, including vulnerable subgroups, and to identify the likely pollutants or mixtures of pollutants. The health outcomes will be daily mortality (1987-94), hospital admissions (1992-94), accident and emergency attendances (1992-94) and general practitioner consultations (1988-94) and these will be analysed by age and selected diagnoses. The pollutants to be investigated will be particulates (including PM_{10} - from 1992), nitrogen dioxide, ozone, sulphur dioxide and carbon monoxide. In addition, the effects of aeroallergens (pollens and fungal spores) will be examined to see if these interact with air pollution in relation to asthma and allergic rhinitis.

The statistical analysis will employ Poisson regression techniques controlling for trends, seasonal and other cycles, temperature, humidity, day of the week and holiday effects and serial correlation.

Methods of quantifying the population impact and health care costs of short-term health effects will be developed using the exposure response relationships obtained from studies of utilisation data.

Author(s):   Dr Raymond Agius

Project Title: The Relationship Between Urban Pollution and Cardio respiratory Ill-Health

Official Address: The University of Edinburgh
                             Department of Public Health Sciences
                             Teviot Place
                             Edinburgh
                             EH8 9AG

Project Abstract: This study aims to investigate the temporal relationships between concentrations of outdoor pollutants, and morbidity and mortality from defined cardiorespiratory diseases using data collected over a much longer time interval than most previously published work. It will relate urban pollution in the city of Edinburgh, notably from particulate matter and sulphur dioxide over 15 years (and for nitrogen dioxide and ozone over about 5 years), to cardiovascular and respiratory mortality (approximately 10,000 cases per annum) and emergency hospital admissions (approximately 10,000 per annum). Utilising extensive data series already collected, it will take account of seasonal, meteorological and pollen effects. By means of modelling it will apply differential exposure estimates across the city. Through indicators derived from the census it aims to resolve confounding due to socioeconomic variables. It will also study possible susceptible groups in a cohort of 1,592 people recruited in 1987/88, who have had measurements of fibrinogen and other risk factors on recruitment. Thus, the influence of a number of cardiorespiratory risk factors on the subsequent relationship between outdoor pollution concentrations and cardiorespiratory ill-health will be determined, in the light of a causal hypothesis linking fine particulate inhalation to the risk of thrombosis.

Author(s):   Dr John Britton

Project Title: Study of the Aetiological Effect of Vehicle Traffic
                        Pollution in the Prevalence and Natural History of
                        Asthma in Nottingham Schoolchildren

Official Address: Nottingham City Hospital
                        Division of Respiratory Medicine
                        Hucknall Road
                        Nottingham
                        NG5 1PB

Project Abstract: This study will investigate the hypothesis that asthma in
                        children is caused or exacerbated by pollutants released
                        by motor vehicles. It will determine the independent
                        effect of exposure to high flow and congestion of
                        traffic and the modifying effects of low birth weight,
                        pre-term birth, exposure to maternal and environmental
                        cigarette smoke, maternal age, infant feeding practice,
                        and socioeconomic status on the occurrence of wheezing,
                        diagnosed asthma and the frequency of symptoms in
                        all Nottingham schoolchildren from age 5 to 16. The
                        study will also establish the current prevalence of
                        asthma and wheezing illness in schoolchildren in
                        Nottingham, and by using data from previous surveys in
                        primary schools in 1985 and 1988, will determine
                        whether the prevalence of asthma is continuing to
                        increase and whether areas with a relatively high
                        increase are also those with the greatest growth in
                        traffic flow. Using longitudinal data from 10,000
                        children first surveyed in 1988, the study will also
                        establish whether incidence or prognosis of wheezing
                        illness in children is independently related to changes in
                        the level of exposure to vehicle traffic pollution.
                        Wheezing history and details of all potential risk factors
                        will be obtained by parental questionnaire from
                        approximately 16,500 and 11,500 primary and
                        secondary schoolchildren, respectively, and traffic flow
                        and congestion data will be obtained by laying flow and
                        speed pneumatic tubes in the vicinity of every school
                        in Nottingham.

Author(s):   Dr Joanne Clough

Project Title: Indoor Pollutants as a Risk Factor for Chronic Respiratory Symptoms in Adolescents

Official Address: University of Medicine, Level G
                       Southampton General Hospital
                       Tremona Road
                       Southampton
                       SO16 6YD

Project Abstract: The aim of this study is to examine the influence of indoor air pollutants on the development of chronic respiratory symptoms in 14-16 year olds. The project will be conducted in two stages and will make use of a cohort of 3,196 children which has been studied since 1987. In the first stage, a detailed postal questionnaire, which will provide information concerning pollutant exposure in the children's home environment, will be sent to all members of the cohort. This, together with data gathered during previous studies on the cohort, will enable us to examine the role played by indoor pollution levels in the inception and persistence of respiratory symptoms in adolescence.

The second stage will involve a longitudinal prospective study of 150 symptomatic children, randomly selected in equal proportions from high and low exposure "risk groups". Diary cards and home monitoring will be employed to assess, over a six month period, respiratory symptoms frequency, severity and duration, and indoor pollutant exposure in these children. Data obtained from this study will enable us to examine the relationship between exposure to high levels of indoor pollutant and respiratory symptoms in adolescents.

Author(s):   Dr David Strachan

Project Title: Chronic Respiratory Health Effects of Cumulative Air
                              Pollution Exposure: A National Birth Cohort Study

Official Address: St George's Hospital Medical School
                              Department of Public Health Sciences
                              Cranmer Terrace
                              London
                              SW17 0RE

Project Abstract: The aims of this study are to investigate relationships of
                              past and current outdoor and indoor air pollution
                              exposure to:

                              (a) lower respiratory illness during childhood;
                              (b) chronic respiratory symptoms in young adults;
                              (c) ventilatory function in young adults;
                              (d) allergic disease in children and young adults;
                              (e) allergic sensitisation in young adults.

                              The study sample will comprise 7,444 members of the
                              British 1958 birth cohort followed up at ages 7, 11, 16
                              23 and 33. A subsample of 1,449, including all those
                              with a history of asthma, wheezy bronchitis or
                              pneumonia and a random sample of the remainder, were
                              examined at age 34-55.

                              Prospectively recorded histories of childhood asthma,
                              wheezy bronchitis, pneumonia, hay fever and eczema:
                              adult reports of cough, phlegm, wheeze, asthma and
                              hay fever: plus, in the subsample of 1,499, lung function
                              and skin prick test responses, will be related to air
                              pollution indices derived for areas of residence at birth,
                              7, 16, 23 and 33. These include measured smoke and
                              sulphur dioxide levels from birth to age 23 and
                              modelled nitrogen dioxide exposure at age 33. Analyses
                              will control for parental smoking in childhood and both
                              active and passive smoking in adult life. Analyses of the
                              1,449 subjects examined will include past and current
                              use of gas for cooking.

Author(s):   Dr S Walters, Professor J G Ayres,
                              Professor R H Harrison

Project Title: Effect of Fine Particulate Air Pollution and Acid Aerosols
                              on   Respiratory   Function   and   Symptoms   in
                              Schoolchildren

Official Address: University of Birmingham
                              The Medical School
                              Edgbaston
                              Birmingham
                              B15 2TT

Project Abstract: The aims of this study are to: (a) discover whether
                              schoolchildren with pre-existing respiratory disease
                              show greater lung function or symptom response to fine
                              particulate air pollution than normal children; and, (b)
                              to determine whether particle size (PM_{10} or $PM_{2.5}$) or
                              composition (acid aerosols) have the greatest impact on
                              respiratory function and symptoms in schoolchildren
                              aged 9-11, living in Birmingham and Sandwell, West
                              Midlands. The study sample will comprise 100 children
                              from participating schools with history of recurrent
                              wheeze in the last year. One hundred and fifty children
                              from the same schools with no history of wheeze
                              diagnosed asthma will also be studied. Schools will
                              come from residential areas representing background
                              levels of pollution, and adjacent to industrial sites and
                              the M6 motorway. The study will take the form of a
                              prospective panel study. Baseline lung function and
                              skin allergen sensitivity will be tested in participating
                              children. They will keep a diary recording twice daily
                              peak expiratory flow and respiratory symptoms for a
                              two month period during the winter and again during
                              the summer. Concurrent measurements of PM_{10}, $PM_{2.5}$,
                              acid aerosols will be made at each site, and background
                              levels of NO_2, SO_2, O_3 and CO will be available from
                              EUN air quality monitoring sites in Birmingham and an
                              equivalent station in Sandwell. Regression coefficients
                              taking account of contemporaneous weather, day
                              of week, linear trend and autocorrelation will be
                              determined for each child for each pollutant using a
                              two-stage modelling method. The prevalence of
                              responsiveness in each group will be determined, as
                              well as the magnitude of response to each measure of
                              particulates and acid aerosols.

Author(s): Professor A Seaton

Project Title: Air Pollution and Cardiovascular Disease: An Investigation of the Relationship Between Particulate Air Pollution and Blood Coagulation Factors

Official Address: University of Aberdeen
Department of Environmental and Occupational Medicine
Medical School
Foresterhill
Aberdeen
AB9 2ZD

Project Abstract: Particulate air pollution has been shown to be associated with increased death rates from heart attack and stroke in older people. The authors have suggested that this may be a consequence of pulmonary inflammation leading to release of mediators of fibrinogen production and consequent increased coagulability of the blood. This might explain both the observed short term and the possible longer term effects of pollution on cardiovascular mortality. This study proposes to investigate this association, using time-series analysis of the inter-relationships of exposures of particulate pollution and clotting factors in the blood. The techniques that have been used to show a relationship between ambient temperature and clotting factors, studying 50 subjects aged 60+ years in both Belfast and Edinburgh over two years will be used. Measures of fibrinogen, factor VII and IL-6 will be undertaken monthly for all subjects, use will be made of routinely recorded local air pollution data, and the study will also make estimates of individuals' exposures to particulate pollution from activities diaries and measurements of the exposures of a sample using portable dust samplers. The study is intended to explain the link between particulate pollution and cardiovascular disease and may indicate the levels of exposure to particles at which such effects might be expected to occur.

Author(s):	Dr Michael Burr
Project Title:	The Effects of Relieving Traffic Congestion on Pollutant Exposure and Respiratory Morbidity
Official Address:	University of Wales College of Medicine Temple of Peace and Health Cathays Park Cardiff CF1 3NW

Project Abstract:

The aims of this project are: (1) to see whether people who are frequently exposed to high levels of vehicle exhaust fumes are more liable than other people to have impairment of respiratory health and to be allergic to aeroallergens; and (2) to see whether respiratory health improves when people cease to be exposed to high levels of air pollution from vehicle exhaust fumes.

The subjects will be residents of villages that at present are subject to heavy traffic congestion which will be diverted around a by-pass to be opened within the next two years. A respiratory survey will be conducted among the residents of uncongested streets with similar types of houses in the same area. About 600 residents in each case will be enlisted. Symptom questionnaires will be completed and health service usage documented. Lung function and skin prick tests will be performed on all subjects over the age of 7 years. Blood will be taken from the adults for IgE measurement. Air pollution will be monitored by air sampling and personal NO_2 and CO badges. A year after the by-pass opens the respiratory survey and measurements of pollutants will be repeated.

Author(s):   Professor J G Ayres, Professor R M Harrison

Project Title: To Assess the Effect of Challenge with Fine and Ultra-Fine Particles on Airway Diameter and on Subsequent Response to Allergen Challenge in Patients with Asthma

Official Address: Birmingham Heartlands Hospital
                              Department of Respiratory Medicine
                              Bordesley Green East
                              Birmingham
                              B9 5SS

Project Abstract: The aims of this project are twofold.

(1)   To determine whether, in mild asthma, exposure to sulphuric acid, ammonium bisulphate (100 and 1000 $\mu g/m^3$, 1 μm MMAD) and ultra-fine carbon black particles (200 nm), alone or in combination, results in changes in ventilation, spirometry and exhaled NO at rest and on exercise. Particle challenge will employ a Vibrating Orifice Aerosol Generator using a closed circuit system. Volume breathed, respiratory rate, FEV_1 and FVC will be measured over four hours using the Spirotrac system with exhaled NO measured at the same time intervals. The presence or absence of a bronchoconstrictor response will impact on the design of the next part of the study.

(2)   To determine whether exposure to sulphuric acid and ammonium bisulphate for 1 hour enhances the bronchoconstrictor effect to grass pollen challenge in grass pollen sensitive subjects with asthma. The acute response (in terms of change in FEV_1) will be the primary outcome variable, but an assessment of the late response will also be made.

Dependent on the results of the above, the effects of carbon black exposure on allergen challenge will be determined. These studies will provide information on the possible permissive effect of particle pollution on allergen challenge, a more subtle and difficult effect to determine than that of single pollutant challenge.

Author(s):	Dr W Keys
Project Title:	Secondary Schoolchildren with Asthma
Official Address:	NFER The Mere Upton Park Slough SL1 2SQ

Project Abstract:

This research will focus on pupils in the age range 11-16 and will seek to investigate three main questions:

(1) How do asthmatic children obtain information on air pollution?

(2) How do they respond to such information?

(3) How could such information be better targeted for this group?

Phase 1 of the research will consist of discussion with doctors involved in the care of children with asthma, representatives of support and pressure groups, and staff and asthmatic pupils in about five schools. A brief review of the relevant literature will also be undertaken. Phase 2 will consist of a questionnaire survey of asthmatic pupils in about 200 schools. This survey will be complemented by a survey of schools designed to elicit information on secondary schools' policies with regard to asthmatic pupils. In Phase 3, a number of group discussions will be held with pupils to explore selected issues in greater depth.

Author(s):   Mr J F Hurley, Professor R M Haines,
                              Professor J G Ayres, Dr A Markandya

Project Title: Towards Assessing and Costing the Health Impacts of
                              Ambient Particulate Air Pollution in the UK

Official Address: Institute of Occupational Medicine
                              8 Roxburgh Place
                              Edinburgh
                              EH8 9SU

Project Abstract: This feasibility study aims to quantify and cost the
                             health effects of particulate air pollution in the UK; and
                             to assess the reliability of the resulting estimates.

                             The strategy uses methodology for quantified risk
                             assessment, with costings, developed/implemented by
                             various participants (ETSU, IOM, Metroeconomica)
                             within recent EC-sponsored studies (EC DGXII). That
                             methodology will now be critically reviewed with other
                             experts, especially for its relevance to UK conditions;
                             and, where practicable, extended and improved.

                             Regarding sample size, type and location, the study
                             does not involve new epidemiological, laboratory,
                             environmental or economic investigations. Rather, it
                             draws on relevant studies, UK and internationally, and
                             existing UK background data.

                             The methods of working are an integrated scheme of
                             assessing and where practicable improving: (1) an
                             epidemiologically-based risk model, with exposure-
                             response relationships, linking ambient particulates with
                             various health endpoints; (2) background UK data on
                             particulate ambient pollution, mortality and morbidity;
                             (3) economic valuation of health effects, principally
                             using the Willingness-to-Pay approach; (4) linkage of
                             these within a Geographical Information System to give
                             estimated impacts and costs. Results will be reported
                             with expert assessment of their reliability.

Author(s):

Professor R J Davies
Dr J D Devalia

Project Title:

The Effect of Combination of Pollutants (NO_2 + O_3) at Different Concentrations and Exposure Times on the Airway Response of Mild Asthmatics to Inhaled Allergen Over a Period of 48 Hours

Official Address:

Department of Respiratory Medicine and Allergy
St Bartholomew's Hospital
London
EC1A 7BE

Project Abstract:

The aims of this project are (1) to study the concentration and time (dose) effects of NO_2, O_3 and their combination on the airway response to inhaled allergen in mild asthmatics and (2) to determine the period of time during which the enhanced airway response is maintained and when it is maximal. Four groups of 13 mild asthmatic volunteers will be investigated in an environmental chamber in a single-blind manner. One group will undergo exposure for 6 hours to air, 200 ppb NO_2, 100 ppb O_3, or a combination of the two pollutants, and then challenged with allergen to determine the dose of allergen required to reduce the FEV_1 by 20% (PD_{20} FEV_1). Another group will undergo 6 hour exposures, once to air and on three separate occasions to a combination of 200 ppb NO_2, + 100 ppb O_3, and challenged with allergen either immediately, 24 hours or 48 hours after exposure, to determine PD_{20} FEV_1. The third and fourth groups will follow the same experimental protocols as groups one and two, except that the mild asthmatics will be exposed to twice the concentration of pollutants (400 ppb NO_2 and 200 ppb O_3) for half the time (3 hours). All volunteers will exercise intermittently during each of the exposures to air and the pollutant gases, which will be randomised and at least two weeks apart.

Author(s):

Dr D Strachan
Professor P Elliott
Mr D Briggs

Project Title:

Relationship of Asthma and Allergic Rhinitis to Local Traffic Density and Ambient Pollution Modelled at a Small Area Level

Official Address:

Department of Public Health Sciences
St George's Hospital Medical School
Cranmer Terrace
London
SW17 ORE

Project Abstract:

The aims of this study are to investigate small area variations in the prevalence and severity of wheezing illness, seasonal allergic rhinitis and allergic sensitisation among children in Sheffield, and their relationship among individuals to estimates of exposure to traffic-related air pollution. 18,203 respondents to a questionnaire survey of asthma, allergic rhinitis and eczema among all 23,084 secondary school children aged 11-16 years in Sheffield during 1991 will be followed. Subsequent studies will include skin prick tests on 727 of these children, and a case-control study of indoor environmental exposures among all 763 children with more severe wheezing, 763 controls who had never wheezed, and 359 children with milder forms of wheezing. Exposure to traffic pollution will be estimated by three different Geographical Information Systems techniques:

(a) proximity of home and school to the nearest main road;
(b) density of traffic flow along major roads within 200 m radius of the home and school;
(c) estimates of pollutant concentrations at place of residence and schooling based on pollution modelling validated against measured NO_2 levels.

Small area variations in symptoms will be assessed by map smoothing techniques and relationships to estimated pollution exposure analysed at the individual level by logistic regression.

Author(s):   Dr C Luczynska
                              Mr J McAughey
                              Dr F Kelly
                              Dr J Sterne
                              Dr D Wheeler
                              Mr J Rice
                              Professor P Burney

Project Title: The Acute Effects of Particulate Air Pollution in Patients with Respiratory Disease (Thamesmead Air Pollution and Respiratory Health Survey)

Official Address: Department of Public Health Medicine
                              UMDS St Thomas' Campus
                              Lambeth Palace Road
                              London
                              SE1 7EH

Project Abstract: The proposed research will study the acute effects of particulate air pollution on a susceptible group of adults with chronic respiratory disease including asthma, from a South London general practice. The study will have two parts: 1) a cohort study of the association between exacerbations of respiratory symptoms and air pollutants, and; 2) a nested case-control study of the relationship between exacerbations and serum specific IgE. A cohort of patients will be recruited and baseline information obtained. This will include measurements of lung function, including peak flow, atopic status and antioxidant status using both serum and nasal lavage fluid, which can be used as a proxy for lung lining fluid levels. Patients who present with exacerbations of respiratory symptoms during the study period will be compared with age and sex-matched members of the cohort who have not presented with symptoms and do not have reduced peak flow at that time. Patients' sera will be used to investigate the allergenicity of particulate air pollution. Levels of PM_{10} and other air pollutants will be measured during the study period and the confounding effects of allergen exposure, infections and smoking will be controlled for. We will investigate whether exacerbation of symptoms can be explained by atopy, sensitisation to any local airborne allergen present, and antioxidant status.

Author(s):   Dr A J Frew
                                    Dr T Sandstrom
                                    Professor S T Holgate

Project Title: Study of Lung Function and Biochemical and Cellular
                                    Consequences of Acute Exposure to Diesel Exhaust in
                                    Normal Asthmatic Subjects

Official Address: Department of Medicine, Centre Block
                                    Southampton General Hospital
                                    Tremona Road
                                    Southampton
                                    SO16 6YD

Project Abstract: Emissions from diesel engines are a major contributing
                                    source to particulate matter along with nitrogen oxides,
                                    formaldehydes and hydrocarbons. Knowledge of the
                                    toxic effects of these pollutants on human lung is
                                    rudimentary and is in urgent need of study as several
                                    epidemiological studies have shown a consistent
                                    association of particulate pollution with increased
                                    mortality and morbidity from respiratory and cardiovas-
                                    cular disease. In this study we will expose 15 normal
                                    and 15 asthmatic volunteers to a controlled concentra-
                                    tion of diluted diesel exhaust and filtered air in a
                                    chamber which has already been developed and
                                    standardised for this study in Sweden. We will then
                                    measure lung function changes and obtain
                                    bronchoalveolar lavage and endobronchial biopsies to
                                    assess the antioxidant defence screen in the lung lining
                                    fluid, mediator release and cellular activation in BAL
                                    fluid. The histology of bronchial mucosa will be
                                    assessed with special reference to cytokine, endothelial
                                    adhesion molecule and leukocyte surface marker
                                    expressions using modern molecular techniques of
                                    mediator measurement, flow cytometry, RT-PCR,
                                    quantitative PCR and immunochemistry that we have
                                    established in our laboratory. This study will give us a
                                    comprehensive picture of proinflammatory effects of
                                    diesel exhaust and help to inform future transport
                                    policy making decisions.

Author(s):   Dr T Pless-Mulloli
                        Ms D Howel
                        Ms J Tate

Project Title: Do Particulates from Opencast Coal Mining Impair
                        Health?

Official Address: Department of Epidemiology and Public Health
                        The Medical School
                        University of Newcastle
                        Newcastle upon Tyne
                        NE2 4HH

Project Abstract: The aim of this project is to test the hypothesis of an
                        association between exposure to dust from soil
                        overburden and coal, and health effects on communities
                        living nearby. The study will compare the health of
                        children aged 1-11 years living close to active opencast
                        coal sites, with controls from otherwise similar control
                        communities.   Five pairs of study and control
                        communities, with approximately 300 children in each,
                        will be selected according to agreed criteria. Suitable
                        sites in the Northern and Yorkshire Region have already
                        been identified. The respiratory morbidity of children
                        will be assessed from self-reported ill health episodes
                        using a questionnaire, a daily respiratory diary and an
                        analysis of GP records. Exposure to particulates will be
                        assessed by real time monitoring of PM_{10} levels and
                        qualitative analysis of filters which will identify the
                        composition and therefore, likely source of particulates.
                        Monitoring and daily diary data will be collected for
                        a six week period in each community; the GP records
                        over one year. The exposure/health relationship will
                        then be analysed.

Appendix 3

Summary of Committee Reports

COMEAP Report: Non-Biological Particles and Health

A3.1   The DH Committee on the Medical Effects of Air Pollutants (COMEAP) reported on 8 November 1995. The Committee's conclusions are summarised below:

There is clear evidence of associations between concentrations of particles similar to those encountered currently in the UK, and changes in a number of indicators of damage to health. These range from changes in lung function through increased symptoms and days of restricted activity to hospital admissions and mortality.

The evidence regarding these associations has been gathered from a range of well conducted epidemiological studies which have mainly been undertaken in the United States (US) and recently in Europe. The consistency of the associations demonstrated by these studies is notable especially as regards mortality, though the reported effects on health of day-to-day variations in concentrations of particles are small in comparison with other uncontrollable factors, eg, seasonal variations or variations in temperature. Considerable coherence of results across health endpoints has also been shown. The few studies which have been conducted in the UK have tended to confirm the findings of studies reported from other countries.

There is no clear evidence that associations with effects on health are restricted to specific types of particles. Epidemiological studies have demonstrated associations between effects on health and particles from a wide range of sources. These include primary emissions from motor vehicles, industrial sources or coal fires and secondary aerosols derived from gaseous emissions, including sulphur dioxide and oxides of nitrogen, from industrial and vehicular sources. In the absence of strong evidence on the relative effects of different particles within the respirable range, it seems reasonable, at present, to base policy on PM_{10} measurements.

The principal question to consider in reviewing the rapidly expanding literature on health effects of suspended particulate matter is whether the statistical associations demonstrated indicate a causal role. There is certainly a remarkable degree of consistency and coherence in the direction and magnitude of findings from a diversity of studies, carried out in different localities in the US and elsewhere, with a range of different health indicators and varying sources of pollution. We consider that the reported associations between levels of particles and effects on health principally reflect a real relationship and not some artefact of technique or the effect of some confounding factor. The indications that the association is likely to be causal are certainly strong.

We conclude, that in terms of protecting public health it would be imprudent not to regard the associations as causal. We also believe that the findings of the epidemiological studies of the acute effects of particles, which have been conducted in the US and elsewhere, can be transferred to the UK, at least in a qualitative sense. However, we consider that there are insufficient UK data available to allow direct extrapolation and reliable estimation of the size of the effects in the UK.

It would be possible, for any health effect of interest, to take a weighted average of the results of well-conducted published studies and apply this to conditions in the UK. [This would usually imply conversion across different measures of particles.] Thus, the relative risk calculated by Schwartz (Schwartz J. Air pollution and daily mortality: a review and meta-analysis. Environ Res 1994;64:36-52.) with regard to effects of particles on mortality was 1.06 (confidence interval [CI]:1.05-1.07) for a 100 $\mu g/m^3$ change in total suspended particles, equivalent to some shortening of life in approximately 1% of daily deaths per 10 $\mu g/m^3$ increase in PM_{10}. Application to the UK of the results even of such structured meta-analyses does not formally take account of uncertainties in extrapolating to different air pollution mixtures (with generally lower concentrations of suspended particles), climate patterns and at-risk populations. Because of these uncertainties, we think it would be unwise to offer a single coefficient with regard to effects on mortality or any other index of ill health. The reader is referred to the tables in Annex 8A to Chapter 8 with the warning that the estimates based on studies reported in these tables are likely to provide only a first approximation to the actual effect. Studies should be undertaken urgently to allow better quantitative predictions to be made.

The only major difficulty in reaching any firmer conclusion about causality is the lack of any established mechanism of action. The mass of suspended particulate matter associated with adverse effects is very small, and while there is evidence relating to acute effects of some components, the fact that in epidemiological studies similar effects have been reported in localities with different types of suspended particulate matter suggests that particles may have a non-specific action. Reported studies indicate a range of effects, from small changes in ventilatory function or exacerbations of asthma through to increases in deaths among the elderly or chronic sick; it does not necessarily follow that the same components would be involved in each effect. The effects have not been explained in terms of the results of conventional inhalation toxicology studies, though few appropriate studies have been reported. It has been suggested, but by no means proven, that ultrafine particles (< 0.05 μm diameter) may play a role. These particles have been shown in recent animal studies to be unexpectedly capable of producing inflammatory reactions in the lungs. Concentrations of such particles would be higher close to sources in the environment because, with time, they would coalesce into larger and more stable forms. They would represent, therefore, only a small proportion of the mass of material measured as PM_{10}, though they would represent a high proportion of the number of particles present.

It is well established from the reported studies that people with pre-existing respiratory and/or cardiac disorders are at most risk of acute effects from exposure to particles. It has been suggested that these effects occur when air pollution aggravates an acute condition such as a respiratory infection, an attack of asthma or a heart attack in people with pre-existing chronic disease. There is no evidence that healthy individuals are likely to experience acute effects on health as a result of exposure to concentrations of particles found in ambient air in the UK.

Evidence regarding the effects of long term exposure to particles on health is even less well developed than that regarding the acute effects. The possibility of confounding in such epidemiological studies is considerable and it is difficult to estimate the exposures of individuals over relevant time periods. Here again, the results of recent US studies reporting associations with mortality, respiratory symptoms and lung function are probably transferable to the UK in a qualitative sense, though the confidence in the accuracy of the predictions is lower than for the acute effects of particles.

Although the evidence is limited, the Committee advises that it would be prudent to consider these associations between long-term exposure to particles and chronic effects as causal.

There is little evidence to show that exposure to atmospheric particles contributes significantly to the burden of cancer in the UK. The presence of genotoxic carcinogens in particles means that such a contribution cannot be ruled out, although it is likely to be very small.

In terms of monitoring levels of particulate air pollution in the UK, we support the continued use of automatic measurement of PM_{10}. Measurement of Black Smoke should also be continued. The need to determine the temporal and spatial mass and number distributions of particle sizes is stressed.

There is a need for research into the effects of particles on health. This research is needed both to improve the predictions of effects which can be made from currently available epidemiological studies and to investigate possible mechanisms of effect. A number of recommendations for research are provided in Appendix 1 of the report.

Conclusions

The Committee considers that the reported associations between daily concentrations of particles and acute effects on health principally reflect a real relationship and not some artefact of technique or the effect of some confounding factor.

In terms of protecting public health it would be imprudent not to regard the demonstrated associations between daily concentrations of particles and acute effects on health as causal.

We find it difficult to reach a firmer conclusion about causality due to the lack of any established mechanism of action.

We believe that the findings of epidemiological studies which have been conducted in the US and elsewhere, of the acute effects of particles, can be transferred to the UK, at least in a qualitative sense.

It is accepted that insufficient UK data are available to establish the reliability of quantitative predictions of the effects of particles upon health in the UK.

We consider that results of recent US studies of the effects of long-term exposure to particles are probably transferable to the UK though confidence in the accuracy of the predictions is lower than with regard to the acute effects. Although the evidence is limited, we advise that it would be prudent to consider these associations as causal.

There is no evidence that healthy individuals are likely to experience acute effects on health as a result of exposure to concentrations of particles found in ambient air in the UK.

COMEAP Report: Asthma and Outdoor Air Pollution

A3.2 COMEAP was asked by DH to advise on possible links between outdoor air pollution and asthma. A sub-group was set up in 1992 to review the evidence and the Committee's report "Asthma and Outdoor Air Pollution" was published in October 1995. The following is the executive summary of the report:

The following terms of reference were provided by the Department:

To advise on:

(a)   The time trends and geographical pattern of asthma in the United Kingdom (UK) and the relationship of air pollution to such trends and patterns.

(b)   The role of air pollution in aggravating existing asthma.

(c)   The possible mechanisms by which air pollution might cause or aggravate asthma.

(d)   Gaps in relevant information.

(e)   Recommendations for further work.

It was recognised that, while this report would focus on outdoor pollution, other forms of air pollution, including that found indoors (where most people spend the majority of their time), or associated with cigarette smoking, might also be relevant to the causation of asthma.

Asthma is a disease of the lungs in which the airways are unusually sensitive to a wide range of stimuli, including inhaled irritants and allergens. This results in obstruction to airflow which is episodic - at least in individuals with early or mild asthma - and which causes symptoms of tightness and wheeziness in the chest.

There has been an increase of about 50% in the prevalence of childhood asthma over the last 30 years, which corresponds to an increase in atopic diseases generally over this time. There has been at least a ten-fold increase in hospital admissions for asthma among children, which may partly reflect changes in medical practice.

Over the period during which asthma has been increasing, emissions of coal smoke and sulphur dioxide have fallen markedly while those of oxides of nitrogen and volatile organic compounds from motor vehicles have increased. During this time emissions of particles from coal smoke have fallen, whilst those from diesel vehicles have increased.

Data on trends and geographical variations in exposure to ozone, nitrogen dioxide and particles from vehicles are limited. The occurrence of ozone episodes in summer has probably increased over this century, but in the 15-

20 years since measurements began, there is no clear trend in annual average concentrations. Annual average nitrogen dioxide concentrations have not increased in large urban centres, although there is some indication of a small increase in other urban areas.

It has been suggested that environmental factors such as air pollution could initiate asthma in previously healthy individuals or provoke or aggravate asthma symptoms in those who are already asthmatic.

While there is laboratory evidence that air pollution could potentially have a role in the initiation of asthma, there is no firm epidemiological or other evidence that this has occurred in the UK or elsewhere.

While there is some epidemiological evidence that air pollution may provoke acute asthma attacks or aggravate existing chronic asthma, the effect, if any, is generally small and the effect of air pollution appears to be relatively unimportant when compared with several other factors (eg, infections and allergens) known to provoke asthma.

There is some laboratory evidence that exposure to the common gaseous pollutants can enhance the response of asthmatic patients to allergens, though the effect does not seem to be large. There is no direct evidence for such an interaction as a result of exposure to outdoor air pollution in the UK.

There is no consistent relationship between trends in the prevalence of asthma and trends in emissions or ambient concentrations of air pollutants. A number of equally, if not more, plausible explanations for the trends in asthma have been hypothesised.

The epidemiological evidence concerning the short term effects of air pollution on asthma indicates that:

(i)   Day-to-day variations in air pollution are likely to have a small effect on the lung function of asthmatic adults and children. In general these changes are unlikely to cause symptoms. However, patients with severe asthma may be more affected because of their lower reserve of lung function. The main effects are observed in the elderly with chronic obstructive lung disease (which includes asthma).

(ii)   Seasonal patterns of asthma bear little or no relationship to those of air pollution.

(iii)   Based on studies from overseas, it is likely that the short-term fluctuations in levels of air pollution currently encountered in the UK are responsible for small changes in the numbers of hospital admissions and accident and emergency attendances for asthma. Limited experience from the UK during well defined air pollution episodes indicates that admissions may be increased by a small amount along with similar increases in admissions for other respiratory diseases.

The epidemiological evidence concerning the geographical distribution of asthma indicates that:

(i)   There is little or no association between the regional distribution of asthma and that of air pollution.

(ii)   Prevalence studies comparing high with low pollution areas have not found consistent associations between outdoor air pollution and asthma prevalence.

(iii)   There is no convincing evidence that asthma is more common in urban areas than in rural areas of the UK. Limited evidence from the UK and other countries suggests a modest relationship between asthma prevalence and local traffic density. The extent to which this is due to air pollution has yet to be determined.

A number of recommendations for further work are made.

Conclusions

As regards the initiation of asthma, most of the available evidence does not support a causative role for outdoor air pollution. (This excludes possible effects of biological pollutants such as pollen and fungal spores.)

As regards worsening of symptoms or provocation of asthmatic attacks, most asthmatic patients should be unaffected by exposure to such levels of non-biological air pollutants as commonly occur in the UK. A small proportion of patients may experience clinically significant effects which may require an increase in medication or attention by a doctor.

Factors other than air pollution are influential with regard to the initiation and provocation of asthma and are much more important than air pollution in both respects.

Asthma has increased in the UK over the past thirty years but this is unlikely to be the result of changes in air pollution.

MAAPE Reports

A3.3 Ozone (1991)

The Advisory Group on the Medical Aspects of Air Pollution Episodes was set up to provide advice to The Chief Medical Officer in accordance with the Terms of Reference:

To consider whether advice about personal protective measures during air pollution episodes should be given by Central Government and, if so, what that advice should be, to whom it should be addressed, and the criteria which should be adopted for issuing any advice.

The Advisory Group was asked to consider these questions with regard to the episodes of elevated ozone concentrations which have occurred in the UK during hot summer weather before progressing to consider other pollutants including sulphur dioxide and nitrogen dioxide.

Reproducible, statistically significant changes in indices of lung function will occur in many people taking vigorous exercise out of doors during episodes of elevated ozone concentration as are sometimes found in parts of the UK during periods of hot summer weather. There is considerable variation in the response to ozone between individuals.

These changes are unlikely to produce irreversible lung damage though individuals who are sensitive to ozone may experience respiratory symptoms, including cough and discomfort on deep inspiration, whilst taking vigorous exercise out of doors.

Although individuals with asthma or other respiratory disorders appear to be no more likely than healthy individuals to be sensitive to ozone, the effects of ozone may be more troublesome in individuals who already have some impairment of lung function. The magnitude of the effects liable to occur does not justify the issuing of general warnings of either the possible effects of ozone or of any need to undertake measures for personal protection during episodes of elevated ozone concentration. Advice should be made available however so that those who are particularly sensitive to ozone will be in a position to take steps to reduce exposure.

Advice should be made available by means of:

(a)   The Meteorological Office being requested to incorporate information on ozone levels in its weather forecasts when peak hourly concentrations in excess of 100 ppb are anticipated on the basis of an analysis of weather conditions and input from the ozone monitoring network.

(b)   The provision of information on the effects of ambient levels of ozone as part of the recorded message currently available on the Air Quality Help Line set up by DOE and the Meteorological Office.

(c)   The issuing each year of appropriate publicity (for example a joint DH/DOE Press Release) drawing attention to the advice available on the Help Line at the beginning of the period when ozone episodes are expected. In this, background information on the health effects of ozone should be made available.

(d)   The issuing of appropriate publicity (for example a joint DH/DOE Press Release) should ozone levels rise, or be expected to rise, above 200 ppb. Should this occur a modified message should be included in the Help Line message.

During episodes of elevated ozone concentration few individuals will need to take steps to reduce their exposure to ozone. However it should be made clear that a reduction in outdoor exercise during the latter part of the afternoon will ameliorate the effects described in paragraphs 1.3 and 1.4 in those who are sensitive. The changes in lung function that might be experienced are not considered to be large enough to warrant other protective measures, such as the wearing of masks.

Much of the information considered in arriving at the conclusions and recommendations listed above originates in the United States. The results of field studies conducted in the US may not be applicable in the UK in every particular as pollutant mixtures differ from country to country. The almost complete lack of data directly applicable to UK conditions continues to be a major problem.

In addition to the lack of field studies and epidemiological studies of the relationships between ozone levels and the prevalence of respiratory diseases in the UK, there is a need for more laboratory based research particularly on the effects of repeated exposure to ozone, interaction between ozone and other pollutants, the effects of ozone on patients with respiratory problems and the repair mechanisms which are called into action after exposure. A number of specific areas where research is needed are identified.

A3.4   Sulphur dioxide, acid aerosols and particulates (1992)

The Advisory Group on the Medical Aspects of Air Pollution Episodes was set up to provide advice to The Chief Medical Officer in accordance with the Terms of Reference:

To consider whether advice about personal protective measures during air pollution episodes should be given by Central Government and, if so, what that advice should be, to whom it should be addressed, and the criteria which should be adopted for the issuing of any advice.

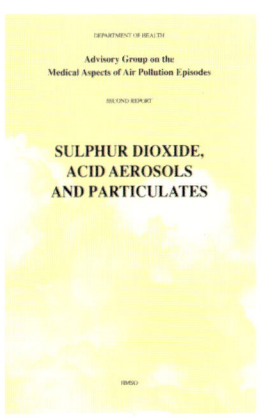

The Advisory Group was asked to consider these questions with regard to the episodes of elevated concentrations of sulphur dioxide which occur in parts of the UK. The Group was also asked to examine the current situation with regard to pollution by particulate material: paying special attention to acid aerosols.

The available evidence indicates that individuals not suffering from respiratory disease will not be affected by episodes of elevated concentrations of sulphur dioxide as occur in the UK. Asthmatic patients are more sensitive to sulphur dioxide. In parts of the UK, levels of sulphur dioxide regularly exceed those at which effects of clinical significance, including tightness of the chest, coughing and wheezing, have been demonstrated in such individuals. These effects are acute and reversible.

The magnitude of the effects liable to occur does not often justify the issuing of warnings but advice (see below) should be made available, particularly to those most likely to be affected:

a. When hourly average concentrations of sulphur dioxide are in the range 125 ppb (357.5 μg/m^3) to 400 ppb (1144 μg/m^3) advice on possible effects should be made available. Asthmatics may experience symptoms including tightness of the chest and coughing and changes in indices of lung function on exposure to concentrations at the upper end of this range. These effects are not expected to be severe nor is there any evidence to suggest that they are indicative of serious or lasting damage to health.
    It is recommended that the advice offered should:

    (i)   Include the information on possible health effects given above.

    (ii) State that these effects may be ameliorated by reducing time spent out of doors and especially by reducing the amount of time spent exercising out of doors.

    (iii) State that there is no evidence to suggest that the wearing of smog masks is necessary.

    (iv) Suggest that those who experience any respiratory effects and those who suffer from chest complaints and who are concerned that they may experience effects, should follow the advice of their doctors.

    (v)   State that asthmatics may need to increase their treatment as advised by their doctors.

b. When hourly average concentrations of sulphur dioxide exceed or are expected to exceed 400 ppb (1144 μg/m^3) a warning should be issued. At these levels many asthmatics may experience significant changes in indices of lung function and symptoms including tightness of the chest, coughing and some breathlessness. These effects may be of sufficient severity to make it advisable for asthmatics to limit exposure, and increase their treatment in consultation, if necessary, with their doctors. Such effects are likely to be mild for the majority of patients and there is no evidence that they have any lasting effect on asthma. Those not suffering from respiratory diseases are not expected to experience any adverse health effects though conditions may be less pleasant than normal.

It is recommended that the advice offered should include the points listed in the paragraph above and in addition:

(i)   State that levels of sulphur dioxide in excess of 400 ppb (1144 μg/m^3) seldom occur in the UK.

(ii) State that the symptoms experienced by asthmatics may be more marked when the weather is also particularly cold.

Advice should be made available by means of:

a.   The Meteorological Office being requested to incorporate information on sulphur dioxide levels in its weather forecasts when peak hourly concentrations of sulphur dioxide exceed or are expected to exceed 125 ppb (357.5 μg/m^3), on the basis of an analysis of weather conditions and the examination of data from the sulphur dioxide monitoring network. It should be made clear which localities are likely to be affected.

b.   The provision of information on the effects of ambient levels of sulphur dioxide as part of the recorded message currently available on the Air Quality Helpline set up by the Department of the Environment (DOE) and the Meteorological Office. Currently the following banding system for describing Air Quality in terms of concentrations of sulphur dioxide (expressed as hourly averages) is used:

Very Good	< 60 ppb	(171.6 μg/m^3)
Good	60-125 ppb	(171.6-357.5 μg/m^3)
Poor	125-500 ppb	(357.5-1430 μg/m^3)
Very Poor	> 500 ppb	(1430 μg/m^3)

We recommend that the advice listed above be provided on the Helpline when Air Quality passes into, or is expected to pass into, the Poor band. It is noted that the point of transition from the Good band to the Poor band is close to the concentration recommended as a 1 hour Guideline Value (122 ppb, 350 μg/m^3) by the World Health Organisation in its Air Quality Guidelines for Europe (1987). It is further recommended that the point of transition from the Poor to Very Poor band be lowered from 500 ppb (1430 μg/m^3) to 400 ppb (1144 μg/m^3) and that a warning be issued with revised advice on the Helpline when Air Quality passes or is expected to pass into the Very Poor band.

c.   The issuing of appropriate publicity (for example a joint DH/DOE Press Release) should sulphur dioxide levels rise or be expected to rise, above 400 ppb (1144 μg/m^3) hourly average. This should draw attention to the Air Quality Helpline.

d.   The issuing each year, at the start of October, of appropriate publicity (for example a joint DH/DOE Press Release) drawing attention to the advice available on the Helpline at the beginning of the period when episodes of elevated concentrations of sulphur dioxide are expected. In this publicity, background information on the health effects of sulphur dioxide should be made available and attention should be drawn to the Air Quality Helpline.

During episodes of elevated sulphur dioxide concentrations in affected localities, only those suffering from respiratory diseases may need to take steps to reduce their exposure. It should be made clear that reducing time spent out of doors and particularly time spent exercising out of doors will ameliorate the effects described in paragraph 1.4. The effects of exposure do not warrant advising that smog masks should be worn.

Suspended Particulate Matter

In comparison with conditions in the UK in the 1950s and 1960s, levels of (non-specific) particulate material are low and are not thought to pose a significant threat to health. However, little monitoring of particulates regarding composition or the mass of particulates per unit volume of air (gravimetric concentration) is undertaken in the UK and thus this conclusion must be regarded as tentative. It is recognised that trace amounts of carcinogens occur in association with airborne particulates. The risk of cancer which might be associated with exposure to particulate pollutants has not been addressed in this Report for several reasons:

a.   The importance of episodic exposure to carcinogens, as compared with long term exposure, is unknown,

b.   detailed information on the nature of carcinogens associated with particulates is lacking,

c.   epidemiological data linking cancer with exposure to ambient levels of air pollutants other than cigarette smoke are very few.

Acid aerosols

Suspended particulate matter includes a small variable acidic component and it is known from laboratory experiments that exposure to acid aerosols may produce changes in indices of lung function and that asthmatic patients may be more sensitive in this regard than other individuals. However, the effects of inhaling acidic aerosols under experimental conditions have been rather small and inconsistent over the short term. There are insufficient data available regarding acid aerosol levels, particularly in urban areas, to allow any assessment of likely effects to be made. We strongly recommend that methods for monitoring ambient levels of acidity (hydrogen ion concentration) be developed and a basic monitoring network be established.

Need for research

Data were found to be lacking in a number of areas and there is a clear need for further research. A number of specific recommendations regarding research are made in the following areas:

a. Biochemical and toxicological mechanisms of action of sulphur dioxide and related pollutants upon the respiratory system.

b. The effects of comparatively low concentrations of sulphur dioxide and related pollutants upon lung function in asthmatics, particularly those with more severe asthma, and others with respiratory disease.

c. Interactions between sulphur dioxide and related pollutants, and also with other pollutants e.g. nitrogen dioxide, ozone.

d. Epidemiological research into the effects of low levels of these pollutants on the prevalence of respiratory disease and on the incidence of asthma attacks.

The need for monitoring of personal exposure to sulphur dioxide and for basic information on levels of acid aerosols in the UK was also identified.

A3.5 Oxides of Nitrogen (1993)

The Advisory Group on the Medical Aspects of Air Pollution Episodes was set up to provide advice to the Chief Medical Officer in accordance with the Terms of Reference:

To consider whether advice about personal protective measures during air pollution episodes should be given by Central Government and, if so, what that advice should be, to whom it should be addressed, and the criteria which should be adopted for the issuing of any advice.

The Advisory Group was asked to consider these questions with regard to the episodes of elevated concentrations of the oxides of nitrogen which occur in parts of the UK. The oxide of nitrogen of most concern in outdoor air is nitrogen dioxide; the Advisory Group has concentrated on this pollutant.

The available evidence indicates that individuals not suffering from respiratory disease will be unaffected by such episodes of elevated concentrations of nitrogen dioxide as occur in the UK. When studied in the laboratory there is no consistent difference in sensitivity to nitrogen dioxide between asthmatic patients and normal individuals. Some recent epidemiological studies however have indicated that persons suffering from respiratory disorders, including asthma, may experience a worsening of their symptoms when ambient levels of nitrogen dioxide and associated pollutants are raised.

The conclusion that few effects are likely to occur at levels of nitrogen dioxide encountered outdoors in the UK leads to the recommendation that health warnings and advice regarding nitrogen dioxide episodes should only be issued in exceptional circumstances.

It is recommended that information on levels of nitrogen dioxide continue to be provided via the telephone helpline service and that when nitrogen dioxide levels enter the "Very Poor" band (over 300 ppb, 564 µg/m³) there should be health advice to those suffering from respiratory disorders, including asthma. Only in the unlikely event that the levels exceed 600 ppb (1128 µg/m³) should there be a warning, possibly accompanied by a press release.

It is recommended that DH(CMO) should write to doctors informing them of the conclusions of this report.

It is not considered necessary to advise individuals to wear protective equipment such as smog masks to defend against the effects of elevated concentrations of nitrogen dioxide.

Need for research

Data were found to be lacking in a number of areas and there is a clear need for further research. A number of specific recommendations regarding research are made in the following areas.

- Biochemical and toxicological mechanisms of action of nitrogen dioxide upon the respiratory system.

- The effects of comparatively low concentrations of, and of repeated exposure to, nitrogen dioxide upon lung function in normal and asthmatic individuals.

- Interactions between nitrogen dioxide and other pollutants, including ozone and sulphur dioxide.

- Epidemiological research into the effects of chronic exposure to nitrogen dioxide such as may occur indoors.

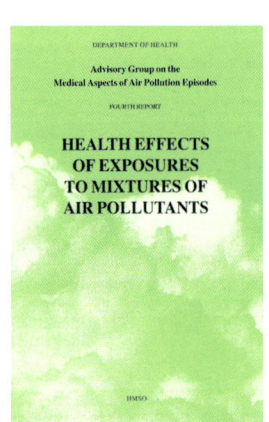

A3.6 Health effects of exposure to mixtures of air pollutants (1995)

The Advisory Group on the Medical Aspects of Air Pollution Episodes was set up to provide advice to the Chief Medical Officer in accordance with the Terms of Reference:

> To consider whether advice about personal protective measures during air pollution episodes should be given by Central Government and, if so, what that advice should be, to whom it should be addressed, and the criteria which should be adopted for the issuing of any advice.

The Advisory Group was asked to consider the effects of exposure to air pollution episodes in which the concentration of more than one pollutant was raised above background levels. In the Advisory Group's previous reports the individual air pollutants have been considered separately.

Predicting the effects of exposure to combinations of pollutants proved to be difficult. The Advisory Group therefore recommends caution when considering the effects of exposure to more than one pollutant and points out that the level of understanding of the mechanisms of possible interactive effects is even less well developed than the understanding of the mechanisms underlying the health effects of individual pollutants.

The Advisory Group identified three common types of air pollution episode:

- Type 1: Summer smog, the pollution mixture with the main, or indicator, pollutant being ozone;
- Type 2: Vehicle smog, the indicator pollutant being oxides of nitrogen;
- Type 3: Winter smog, the indicator pollutant being sulphur dioxide, with a contribution from oxides of nitrogen.

Elevated concentrations of particles may occur during all three types of air pollution episode.

A main concern in this report has been the question of whether the biological effects of mixtures of air pollutants differ from those of the individual pollutants in isolation. Potentiation of damage (synergism) has been seen following exposure of animals to some combinations of pollutants. However, there is a paucity of information on the toxicological effects of complex mixtures of pollutants, particularly those that relate to the three main types of pollution episode seen in the United Kingdom.

Pollen and fungal spores, or their fragments, are common aeroallergens during the period between the late spring and the early autumn. They are responsible for attacks of hay fever or deterioration in asthma in sensitised individuals. Elevated aeroallergen levels may occur in conjunction with Type 1 air pollution episodes. A limited number of laboratory studies, in which human subjects have been exposed to relatively high concentrations of air pollutants, have demonstrated an enhanced reaction to aeroallergens.

In laboratory studies of human subjects exposed to combinations of air pollutants the results have been variable. There has been no clear evidence of synergism between pollutants at the concentrations studied, although interactions between air pollutants may have been underestimated in such studies due to the small size of many studies, limited durations of exposure and the confounding effects of prior exposure to ambient pollutants.

It is probable that all three main types of pollution mixtures encountered during air pollution episodes in the United Kingdom could cause small mean reductions in lung function in normal individuals. The evidence for this is probably strongest for ozone related pollution, where reduction in lung function and development of symptoms is more likely in those who exercise. There is no evidence that any of the three types of episode commonly seen in the United Kingdom cause symptoms or adverse health effects in people who are otherwise well.

It is likely that episodes of air pollution occurring in the United Kingdom produce adverse health effects in some persons with chronic respiratory disease. There is also evidence that some individuals with asthma might experience some deterioration in their condition, although the majority have shown little effect in most studies.

The elderly, especially those with chronic cardiopulmonary disorders, are identified as being at increased risk of deterioration during air pollution episodes. The degree of increased risk incurred by these groups is difficult to quantify.

It might be expected that patients with asthma would be especially vulnerable to air pollution. At levels encountered in the United Kingdom this has proved difficult to demonstrate convincingly, except possibly in those with severe asthma. It is possible that asthmatic patients might compensate for any effects of air pollutants by increasing their medication.

Evidence relating to the long term health effects of exposure to such episodes of air pollution as occur now in the United Kingdom is not available.

It is recommended that information on air pollution episodes continue to be provided to the public. More detailed information about the type of air pollution episode expected, the indicator pollutant and any associated pollutants, geographical location, severity and its likely duration should be provided in forecasts.

It is recommended that the banding system nomenclature suggested in previous reports of the Advisory Group be adopted and applied to episodes of air pollution in which levels of pollutants are elevated. If it were to be predicted that the concentration of a pollutant would pass into the "Very Poor" band then the episode should be characterised as "Severe".

The Advisory Group identified criteria that needed to be fulfilled if health advice provided to the public is to be beneficial. These included: the ability to predict an episode sufficiently far in advance, the ability to target advice to groups at risk, and the existence of effective interventions for those at risk.

It was recognised that the identification of groups at risk and the understanding of effective interventions for such groups were each subject to some uncertainty. The Advisory Group therefore recommended that this topic be kept under review by the Committee on the Medical Effects of Air Pollutants.

The Advisory Group considered the medical advice which should be provided during air pollution episodes and recommended that the following text should be used on the FREEPHONE Air Quality Helpline:

'The following advice on health applies only on days when air quality is "Poor" or "Very Poor". If you have a chest problem, such as asthma or bronchitis, you may experience a worsening of your symptoms on these days. If this is the case it might help to increase your treatment temporarily — you should talk to your doctor about the options.'

The use of smog masks during air pollution episodes is not recommended on current evidence.

It is recommended that the findings of this report be drawn to the attention of doctors.

A number of recommendations for further research are made; these are detailed in Appendix 1.

EPAQS recommendations to date

A3.7   Benzene (February 1994)

"The panel recommend an Air Quality Standard for benzene in the United Kingdom of 5 ppb as a running annual average. We also recommend that this Standard be reduced to the lower level of 1 ppb running annual average, and that the Government set a target date by which this be achieved.

These recommendations are intended to reduce the overall levels of exposure of the population to benzene such that ambient air is no longer the main source of individual exposure, even for non-smokers."

Since benzene is a genotoxic carcinogen, and is, therefore, prudently regarded as having no safe level, EPAQS has recommended a lower "target" to be reached at some point in the future.

A3.8   Ozone (May 1994)

"The panel recommend an Air Quality Standard for ozone in the United Kingdom of 50 ppb as a running 8-hour average.
This recommendation is intended to reduce the exposure of the population, including individuals who may be particularly sensitive, to levels at which harmful effects are unlikely to occur.

We recognise that this is an ambitious Standard which will need international action to be achieved. In order to monitor progress we recommend that ozone monitoring data in the UK are reported in terms of the number of days on which the Standard is exceeded at any one site per year. We note from past evidence that if the 50 ppb 8-hour running average is exceeded on less than 10 days per year, then the maximum 8-hour concentration at that site is unlikely to exceed 100 ppb, a concentration at which effects in healthy individuals have been clearly demonstrated"

This recommended standard was exceeded at all monitoring sites during 1990 (a year of frequent photochemical activity), for 6 to 80 days. Exceedences were also widespread during 1994 and it seems likely that there will be many exceedences during 1995.

A3.9   1,3-Butadiene (December 1994)

"The Panel recommend an Air Quality Standard for 1,3 butadiene in the United Kingdom of 1 ppb measured as a running annual average. We also recommend that this Standard, and the need for a target Standard, be reviewed after a period of, at most, five years in the light of any additional human data and of the experience of improved pollution control.   In the interim the Panel recommend that, as a precautionary measure, the Government's programme of air pollution controls should aim to ensure that, at any one site, the Standard is not exceeded and that there is an overall decline in concentrations.

These recommendations are intended to reduce the cumulative lifetime exposure of the United Kingdom population to 1,3 butadiene"

A3.10   Carbon monoxide (December 1994)

"The Panel recommend an Air Quality Standard for carbon monoxide in the United Kingdom of 10 ppm, measured as a running 8 hour average.

This recommendation is intended to limit the exposure of the population, including individuals who may be particularly susceptible, to levels at which harm is unlikely to occur. The Panel note however that, in general, smokers will have elevated levels of carboxyhaemoglobin and will not be influenced by additional exposure to carbon monoxide in ambient air".

A3.11   Sulphur dioxide (September 1995)

"The panel recommend an Air Quality Standard for sulphur dioxide in the United Kingdom of 100 ppb, measured over a 15 minute averaging period.

This recommendation is intended to reduce the exposure of the population, including individuals who may be particularly susceptible, to levels of sulphur dioxide at which harmful effects are unlikely to occur. Although the exposure studies discussed earlier have primarily investigated the effect of sulphur dioxide on people suffering from asthma, it is likely that similar effects may be observed in patients with other chronic lung diseases."

Under the Second Sulphur Protocol of the Convention on Long-Range Transport of Air Pollution, signed in 1994, the UK is committed to a reduction in national sulphur emissions of 50% by 2000, 70% by 2005 and 80% by 2010 compared to a 1980 base. Further consideration is being given to whether these reductions will allow the EPAQS recommendations to be met.

A3.12 *Particles (November 1995)*

"The Panel recommend an Air Quality Standard for PM_{10} in the United Kingdom of 50 µg/m^3 measured as a 24-hour running average. We also recommend that this Standard be reviewed after a period of, at most, five years in the light of any new data. We further recommend that the Government's programme of air pollution controls should aim to ensure that there is a decline in both peak and annual average concentrations of PM_{10}.

These recommendations taken together are intended to reduce the levels of exposure of the population to concentrations at which effects on health will become progressively less easy to detect and, eventually, unimportant in public health terms. It is intended that techniques for monitoring the 24-hour standard should be consistent with the PM_{10} measurements made by the Department of the Environment's Automatic Urban Network".

A3.13 *Nitrogen dioxide (December 1996)*

The panel recommend an Air Quality Standard for nitrogen dioxide in the United Kingdom of 150 ppb measured as an hourly average.

The panel consider that a longer-term Standard is also desirable, but believe that there is as yet insufficient evidence from which to determine an appropriate value. We recommend that this be reconsidered within the next three years and, in the meantime, also recommend a strategy of reduction in the annual average concentration of nitrogen dioxide.

These recommendations taken together are intended to reduce the levels of exposure of the population to concentrations in the outdoor air at which acute health effects will be extremely uncommon in susceptible people and at which more subtle effects on the public health will become progressively more difficult to detect. It is intended that techniques for monitoring the Standard should be consistent with the measurements made by the Department of the Environment in its national automatic nitrogen dioxide monitoring networks.

Appendix 4

COMMITTEE ON THE MEDICAL EFFECTS OF AIR POLLUTANTS STATEMENT ON BANDING OF AIR QUALITY[1]

January 9th 1997

1. The Government is proposing to revise the system of Air Quality Banding used since 1990 to take into account the new Air Quality Standards, recommended by the DOE Expert Panel on Air Quality Standards (EPAQS), outlined in the Air Quality Strategy for the UK and a subsequent separate consultation on nitrogen dioxide. The proposed revised system, like the old, is intended to provide guidance as to the effects of air pollutants on health and would be linked with health advice provided via the Air Quality Helpline. It is proposed that revision of the thresholds should draw on advice from the Department of Health and its Committee on the Medical Effects of Air Pollutants (COMEAP).

2. The current system describes air quality in terms of concentrations of ozone, nitrogen dioxide and sulphur dioxide. It is proposed that these bands are revised and, in addition, a system of bands to describe air quality in terms of concentrations of particulate matter and carbon monoxide be added. The system would thus be more comprehensive than formerly and should provide better insight into the possible effects of exposure to air pollution. The bands for each pollutant, and how they were determined, are discussed briefly below.

Ozone

3. The Air Quality Standard (AQS) for ozone is expressed as an eight hour average concentration: 50 parts per billion (ppb). This standard includes a margin of safety below the lowest level at which significant effects on health have been described and thus when concentrations of ozone are less than 50ppb (8 hour average) air quality will be described as "very good". The European Union has accepted a Directive on ozone which specifies two higher concentrations: 90 and 180ppb, 1 hour average concentrations, at which information must be provided to the public. Thus these concentrations have been incorporated into UK legislation and the following bands are defined:

    <50ppb(running 8 hour average)-89ppb(1 hour average): "generally satisfactory/moderate air quality"

    90 - 179 ppb (1 hour average): "poor" air quality

    greater than 180ppb (1 hour average): "very poor" air quality.

4. At concentrations of ozone of less than 90ppb it is very unlikely that anyone will notice any adverse effects though effects are detectable at a population level. As concentrations rise towards 180ppb some individuals, particularly those exercising out of doors, may experience eye irritation, coughing and discomfort on breathing deeply. At more than 180ppb these effects may become more severe. Individuals suffering from asthma and other respiratory disorders associated with a reduction in respiratory reserve, may experience earlier and more marked effects.

[1]   The banding proposals are currently under review by the present Administration (August 1997)

Sulphur dioxide

5.   Sulphur dioxide is a respiratory irritant and causes tightening of the airways when inhaled at high concentrations. Those suffering from asthma are significantly more sensitive to sulphur dioxide than other people. This special sensitivity was taken carefully into account in devising the standard recommended by EPAQS, 100ppb (15 minute average), and in devising the following bands.

> less than 100ppb (15 minute average): "very good" air quality. At these concentrations it is most unlikely that anyone, even those suffering from asthma would experience any adverse effects. The figure of 100ppb thus includes a safety margin.

> 100 - 200ppb (15 minute average): "generally satisfactory/moderate" air quality. There is little evidence to suggest that those suffering from asthma would be significantly affected by exposure to concentrations of sulphur dioxide of less than 200ppb. This figure was accepted by EPAQS as the lowest level at which clear though rather small effects had been described.

> 200 - 400ppb (15 minute average): "poor" air quality. The World Health Organisation has suggested that exposure to 400ppb sulphur dioxide may lead to significant narrowing of the airways in those suffering from asthma. For most people the effects expected would not be large though some individuals may be clinically affected. The effects would be reversed by use of the "reliever inhalers" used by those suffering from asthma. Exposure to such concentrations may add to the effects of exposure to other pollutants and allergens and thus asthmatics should be warned that they may need to increase their medication.

> greater than 400ppb (15 minute average): "very poor" air quality. As concentrations rise above 400ppb then more asthmatic individuals may experience adverse effects and should be encouraged to ensure that they have an adequate supply of their "reliever inhaler". At any concentrations likely to be experienced in the UK it is very unlikely that normal individuals will experience any adverse effects.

Nitrogen dioxide

6.   Nitrogen dioxide is a common air pollutant in urban areas and indoors but studies of its effects on either those suffering from asthma or other individuals are more difficult to interpret than those of sulphur dioxide. Evidence of effects at lower levels (200-300ppb) is inconsistent. Many studies show no effects and, in those which do the effects described are generally very small and likely to be insignificant.These studies were examined closely by EPAQS in recommending a standard of 150pPb (1 hour average).

> less than 150ppb (1 hour average): "very good" air quality. At these concentrations it is very unlikely that anyone will experience any adverse effects.

> 150 - 300ppb (1 hour average): "generally satisfactory/moderate" air quality. Studies of volunteers, including those with asthma, exposed to

concentrations of up to 300ppb for one hour do not provide convincing evidence that significant effects on health are likely. Some increase in the response of the lung to substances which produce narrowing of the airways have been recorded on exposure to nitrogen dioxide at these concentrations. Again the studies are inconsistent and the effects are small.

300 - 400 ppb (1 hour average): "poor air quality". Around 300ppb a few studies have shown small direct effects on indices of lung function. In addition, there is evidence from epidemiological studies of the effects of mixtures of pollutants characterised by concentrations of nitrogen dioxide in this range that adverse effects on health may occur. Should effects occur then those with pre-existing disease of the heart or lungs would be likely to be most at risk: see below.

more than 400ppb (1 hour average): "very poor" air quality. At these concentrations epidemiological studies have provided evidence of effects. These included increased admissions to hospital and consultations of General Practitioners. The air pollution episode experienced in London in 1991 was characterised by an increase in concentrations of both nitrogen dioxide and particulate matter and though a possible effect of nitrogen dioxide cannot be excluded, it is not clear which pollutant was responsible for the adverse effects on health. Those suffering from long standing diseases of the heart and lungs should be aware that their condition may worsen as concentrations of nitrogen dioxide move into the "very poor" band. Individuals suffering from asthma do not appear to be at such increased risk on exposure to nitrogen dioxide as they are on exposure to sulphur dioxide.

Particulate matter

7. Particulate matter is monitored in the UK as PM_{10}: ie particles generally less than 10 microns in diameter. A large number of epidemiological studies have shown that day to day variations in concentrations of particles are associated with adverse effects on health. These include increased daily deaths, increased admissions to hospital of patients suffering from heart and lung disorders, and a worsening of the condition of those with asthma. This evidence was reviewed in detail by COMEAP in the report "Non-Biological Particles and Health" published by DH in 1995.

8. A remarkable feature of the evidence is that even at low concentrations of particles effects remain. Of course, as concentrations fall so effects decrease. In a specially commissioned study of the effects of air pollution in Birmingham, a city with a population of 1 million, it was shown that on a day when the concentration of particles rose from the annual average concentration of $25\mu g/m^3$ to $50\mu g/m^3$ (24 hour average) then one more admission to hospital for treatment of respiratory diseases might be expected. $50\mu g/m^3$, 24 hour average concentration, was accepted by EPAQS as a lowest effect level and was recommended as the EPAQS air quality standard. EPAQS also advised that efforts should be made to reduce annual average concentrations of particles in the UK.

9. Because of the continuous relationship between concentrations of particles and effects on health a different approach to devising bands of air quality has been advised by COMEAP. The following break points between bands were agreed: 50, 75 and $100\mu g/m^3$. Thus the bands are:

less than 50µg/m³ (running 24 hour average): "very good" air quality

50-75µg/m³ (running 24 hour average) : "generally satisfactory/moderate" air quality

75-100µg/m³ (running 24 hour average): "poor" air quality

more than 100µg/m³: "very poor" air quality.

10.   It is accepted that these figures are in a sense arbitrary in that they describe increments of effect along a continuum. As concentrations enter the "very poor" band those suffering from disease of the heart and lungs may experience a worsening of their symptoms and should if necessary, consult their doctors. Such concentrations occur fairly frequently in the UK but the great majority of people will experience no adverse effects at all. It is fair to say that our knowledge of the effects of particulate matter on health is evolving rapidly and this advice and the bands defined above may require revision in the light of new evidence.

Carbon monoxide

11.   Carbon monoxide is perhaps the best understood of the pollutants discussed here. The most significant exposure to carbon monoxide occurs in the general population as a result of cigarette smoking. Carbon monoxide interferes with the transport of oxygen by the blood and at high concentrations produces unconsciousness and death. Some 60 accidental deaths occur in the UK each year as a result of exposure to carbon monoxide indoors.

12.   Concentrations of carbon monoxide in the air are related to concentrations of carbon monoxide in the blood in a well understood and predictable way. In describing levels of exposure, it is usual to speak in terms of that percentage of haemoglobin, the essential oxygen-transporting protein of the blood, which is saturated with carbon monoxide. The effects of different percentage saturations have been studied in volunteers and EPAQS reviewed the results of these investigations closely in recommending a standard of 10ppm (8 hour average concentration). At this concentration the blood would reach a saturation of less than 2% and effects on health would be unlikely even in those suffering from heart disease. However, recent epidemiological studies have reported associations between outdoor concentrations of carbon monoxide and admissions to hospital for treatment of heart disease and this new evidence needs to be kept under review.

13.   Studies of the effects of higher saturations have been carried out and these form the basis of the following banding system.

less than 10ppm (running 8 hour average): "very good" air quality.

10-15ppm (running 8 hour average): "generally satisfactory/moderate" air quality. This level of exposure leads to a saturation of haemoglobin of about 2.5%. At this level there is some evidence to show that those with angina and other heart diseases may experience a more rapid onset of chest pain on exercise.

15-20ppm (running 8 hour average): "poor" air quality. Exposure to 20ppm of carbon monoxide leads to a saturation of 4.5%. At these saturations it has been shown that there is a reduction in peak exercise capacity of healthy subjects and a reduction in the time needed for anginal pain to appear on exercise in those with heart disease. Whether such patients are likely to take sufficient exercise to reveal these effects is questionable but they should be aware that such effects may occur.

more than 20ppm (running 8 hour average): "very poor" air quality. As concentrations of carbon monoxide rise above 20ppm then so the percentage saturation of haemoglobin will increase. Effects on those with heart disease become more likely. It should be recalled that sufficient outdoor exposure to reach these levels of saturation is unlikely.

Proposed Descriptors, Thresholds, and Associated Action for Five Air Pollutants

Description" ... air quality	"Standard Threshold" generally satisfactory/moderate	"Information Threshold"	"Alert Threshold"
	very good	poor	very poor
Sulphur Dioxide	100ppb/15 minute	200ppb/15 minute	400ppb/15 minute
Ozone	50ppb/8 hour running	90ppb/hour	180ppb/hour
Carbon Monoxide	10ppm/8 hour running	15ppm/8 hour running	20ppm/8 hour running
Nitrogen Dioxide	150ppb/hourly	300 ppb/hourly	400ppb/hourly
Fine particles (PM_{10})	50µgm^{-3}/24 hour running	75µgm^{-3}/24 hour running	100µgm^{-3}/24 hour running
Effects	Harmful effects unlikely to occur, even in sensitive groups. There could be a small risk of effects in sensitive individuals and/or disamenity may begin to occur.	Some adverse effects in sensitive groups may occur and disamenity would increasingly be experienced.	Risk of more serious adverse health effects, not necessarily confined to sensitive groups. Serious disamenity may occur.
Action		Provision of relevant advice to those who may be affected.	Provision of advice more generally. Additional action may be necessary.

Available from Air and Environmental Quality Division, The Department of the Environment, Room B3.56, Romney House, 43 Marsham Street, London SW1P 3PY

Bibliography

General

World Health Organization. Air Quality Guidelines for Europe. WHO Regional Publications, European Series No 21. Copenhagen: World Health Organization, 1987.

World Health Organization. Update and Revision of the Air Quality Guidelines for Europe. Meeting of the Working Group "Classical" Air Pollutants, Bilthoven, The Netherlands 11-14 October 1994. WHO Regional Office for Europe. Copenhagen: World Health Organization, 1995.

Department of the Environment. The Scottish Office. The United Kingdom National Air Quality Strategy. Consultation Draft. London: Department of the Environment, 1996.

Department of Health. Advisory Group on the Medical Aspects of Air Pollution Episodes. Fourth Report. Health Effects of Exposure to Mixtures of Air Pollutants. London: HMSO, 1995.

Department of Health. Committee on the Medical Effects of Air Pollutants. Asthma and Outdoor Air Pollution. London: HMSO, 1995.

Ozone

Commission of the European Communities. 92/72/EEC. Air Pollution by Ozone.

Department of Health. Advisory Group on the Medical Aspects of Air Pollution Episodes. First Report. Ozone. London: HMSO, 1991.

Department of the Environment. Expert Panel on Air Quality Standards. Ozone. London: HMSO, 1994.

Sulphur dioxide

Department of Health. Advisory Group on the Medical Aspects of Air Pollution Episodes. Second Report. Sulphur Dioxide, Acid Aerosols and Particulates. London: HMSO, 1992.

Department of the Environment. Expert Panel on Air Quality Standards. Sulphur Dioxide. London: HMSO, 1995.

Nitrogen dioxide

Department of Health. Advisory Group on the Medical Aspects of Air Pollution Episodes. Third Report. Oxides of Nitrogen. London: HMSO, 1993.

Department of the Environment. Expert Panel on Air Quality Standards. Nitrogen Dioxide. London: HMSO, 1996.

Anderson HR, Limb ES, Bland JM, Ponce de Leon A, Strachan DP, Bower JS. The health effects of an air pollution episode in London, December 1991. London: St George's Hospital Medical School, 1994.

Particles

Department of Health. Committee on the Medical Effects of Air Pollutants. Non-Biological Particles and Health. London: HMSO, 1995.

Department of the Environment. Expert Panel on Air Quality Standards. Particles. London: HMSO, 1995.

Anderson HR, Limb ES, Bland JM, Ponce de Leon A, Strachan DP, Bower JS. The health effects of an air pollution episode in London, December 1991. London: St George's Hospital Medical School, 1994.

Carbon monoxide

Department of Health. Expert Panel on Air Quality Standards. Carbon Monoxide. London: HMSO, 1995.

World Health Organization. Update and Revision of the Air Quality Guidelines for Europe. Meeting of the Working Group "Classical" Air Pollutants, Bilthoven, The Netherlands 11-14 October 1994. WHO Regional Office for Europe. Copenhagen: World Health Organization, 1995.

Appendix 5

Glossary of Terms and Abbreviations

Many specialised terms are defined in the body of the report. The glossary includes other terms which may be unfamiliar to readers. Terms are defined with regard to the sense in which they are used when dealing with air pollutants.

Aerodynamic diameter	A term used to describe the size of an airborne particle. A particle of aerodynamic diameter of 2 mm will settle under the influence of gravity at the same rate as a spherical particle with a density of 1 and a diameter of 2 mm. Thus, a very low density particle will have a real diameter greater than its aerodynamic diameter.
Aerosol	A stable suspension of particles in a liquid or a gas.
Allergens	Substances capable of provoking a response by the immune system. Pollen, for example, is a source of allergens and may cause and provoke asthma or hay fever.
Alveolar region of the lung	The part of the lung where gas exchange between the blood and the air takes place. The air is contained in small spaces (alveoli) each about 0.3 mm in diameter.
Anaemia	A disease involving a reduction in the number of red cells in the blood or the production of red cells containing inadequate amounts of haemoglobin. Often caused by a deficiency of iron.
Asthma	A disease of the airways of the lungs in which the airways become inflamed and prone to become narrowed too much and too easily in response to provoking stimuli including allergens and irritating chemicals.
Benign	A term used to describe a type of tumour which grows locally but does not spread through the body via the blood or lymphatic system.
Bronchial mucosa	The lining of the airways or bronchi. This lining bears tiny hair-like structures, cilia on its surface and the regular movement of these moves mucus and inhaled materials up the airways and out of the lungs.

Bronchiolitis — Inflammation of the small airways (bronchioles) of the lungs.

Bronchoconstriction — Narrowing of the airways, ie, of the bronchi. Bronchoconstriction occurs during asthma attacks and can be reversed by use of a "reliever inhaler". Bronchoconstriction may be prevented by use of a "preventer inhaler".

Carboxyhaemoglobin — Carbon monoxide combines with haemoglobin to produce carboxyhaemoglobin. The capacity of the blood to carry oxygen is thereby reduced.

Carcinogen — A compound which can cause cancer.

Catalyst — A substance which increases the rate of a chemical reaction. Platinum in catalytic converters, for example, catalyses the conversion of carbon monoxide to carbon dioxide.

Chamber studies — Studies involving the exposure of volunteers to controlled concentrations of gases or aerosols.

Chronic bronchitis — A chronic inflammatory disease of the lungs commonly caused by cigarette smoking.

Cohort studies — Studies in which a group or "cohort" of people are followed over time to see whether they develop a disease in response to exposure to the factor of interest.

Cytokines — A large group of inflammatory peptide or protein mediators.

DNA — Desoxyribosenucleic Acid. The material found in the nucleus of cells which carries the genetic code.

Emphysema — A chronic lung disease in which destruction of areas of the lung takes place. Cigarette smoking is a common cause.

Enzymes — Chemicals, usually proteins, which act as catalysts for chemical reactions in the body.

Epidemiology — The study of diseases or effects of factors such as pollutants in populations.

ETS — Environmental tobacco smoke. Smoke generated directly by burning tobacco or exhaled into air by smokers.

Ferrous	Containing a particular type of iron (Fe^{2+}).
FEV_1	The volume of air which can be expired during the first second of a forced expiration, ie, during blowing out as hard as possible.
Fibrinogen	A protein which plays an essential role in blood clotting during which it is converted into strands of fibrin.
Genotoxic	A term used to describe carcinogens which act directly, or after transformation in the body, on the genetic material (DNA) of cells.
Haemoglobin	The protein found in red blood cells which transports oxygen.
Heart failure	Failure of the heart to pump an adequate amount of blood. Heart failure may be produced by myocardial infarction.
Immune system	The defensive system of the body which deals with infections and also destroys cells which have undergone malignant change. Diseases such as asthma and some forms of arthritis involve the immune system.
Immunocytochemistry	A technique used in the microscopic study of cells and tissues which involves the use of antibodies to identify or locate particular chemicals.
Leucocytopenia	A reduction in the number of white blood cells (leucocytes).
Leukaemia	A disease in which the number of white cells in the blood is increased due to a cancerous multiplication of immature white cells.
Lung function tests	Measurements designed to investigate lung functions, eg, measurements of lung volume or the maximum rate of air flow through the airways.
Lymphatic system	The system of vessels and lymph nodes which is involved in returning tissue fluid to the blood stream and in the production of antibodies.
Malignant	A term used to describe a type of tumour which spreads through the body by the blood or lymphatic systems and which may invade other tissues aggressively.

Mediator	A term used to describe chemicals produced in the body which provoke a specific response, eg, inflammatory mediators provoke inflammation.
Mesothelioma	A rare malignant tumour of the surface of the lung caused usually by exposure to asbestos fibres.
Meta-analysis	A statistical technique which allows the results of a number of studies to be combined and a type of average result to be derived. The technique allows studies to be given different weightings depending, often, upon their size and quality of design.
Metallo-enzymes	An enzyme containing metal atoms in its molecular structure. Zinc is a common component.
MMAD	Mass Median Aerodynamic Diameter. A term used to characterise the distribution of particle sizes in an aerosol. If the MMAD of an aerosol is 1 mm then this means that half the total mass of material in the aerosol can be found in particles of aerodynamic diameter greater than 1 mm and half in particles of aerodynamic diameter of less than 1mm.
Morbidity	Illness.
Mortality	The mortality rate is the death rate.
Myocardial infarction	The myocardium is the muscular tissue of the heart. If the blood supply to the myocardium is blocked areas of the muscle undergo infarction, ie, damage due to a lack of oxygen. An acute myocardial infarction is a heart attack.
Neurotoxic	Damaging to the nervous system.
nm	Nanometre: one millionth of a mm.
Organic compounds	Chemicals containing carbon, eg, benzene or 1,3-butadiene.
PAH compounds	Polycyclic aromatic hydrocarbons. Chemicals containing carbon atoms arranged in several rings.
Pancytopenia	A reduction in the number of all cell types in the blood.
Pharynx	The anatomical region at the rear of the nose and mouth but above the larynx and oesophagus.

Photochemistry	Chemical reactions brought about by the action of ultraviolet light.
Protocol	A plan for an experiment or study.
Regression analysis	A statistical technique designed to characterise the relationship between changes in some factor (eg, concentration of a pollutant) and a biological response (eg, narrowing of the airways).
SIDS	Sudden Infant Death Syndrome. A term used to describe sudden and difficult-to-explain deaths in babies.
Smog	A term coined in the early 20th Century to describe the mixture of smoke and fog then common in the UK.
Susceptible group	A group of people who as a result of genetic predisposition, illness or unusual exposure are more affected by chemicals than other people.
Synergistic effects	If a person is exposed to two chemicals and the resulting effects are greater than the sum of the effects which would occur if he was exposed to the chemicals singly, then the chemicals are said to act synergistically.
Thrombocytopenia	A reduction in the number of platelets (thrombocytes) in the blood. Platelets are important in blood clotting.
Ultraviolet light	Short wave-length radiation which is just outside the visible spectrum and which causes photochemical reactions.
mm	Micrometre: one thousandth of a mm.

Concentration Units and Conversion Factors

Concentrations of air pollutants are expressed in two ways, either as the mass of pollutant in a given volume of air (usually expressed as micrograms per cubic metre or $\mu g/m^3$) or as the ratio of the volume of the gaseous pollutant (expressed as if pure) to the volume of air in which the pollutant is contained (usually expressed as a volume mixing ratio or parts per million, ppm, or parts per billion, ppb).

The mass concentration as expressed above will be dependent on the ambient temperature and pressure and ideally these should be specified each time a concentration is measured as a mass/volume. The variation is discussed below and although not large may not be negligible where large variations in temperature and pressure occur.

The volume mixing ratio is independent of temperature and pressure, if ideal gas behaviour is assumed.

The relationship between the two sets of units can be expressed as follows:

where:
$$\mu g/m^3 = ppb \; X \; \frac{molecular \; weight}{molecular \; volume}$$

where: $molecular \; volume = 22.41 \; X \; \dfrac{T}{273} \; X \; \dfrac{1013}{P}$

where T is the ambient temperature (°K) and P is the atmospheric pressure (in millibars). Conversion factors for some common gaseous pollutants are given in the Table below for 20°C and 0°C and 1013 mb pressure. Pollutants which are present in particulate form in the atmosphere such as sulphates are normally only expressed in mass/volume units.

Pollutant	Molecular weight	To convert			
		ppb to µg/m³		µg/m³ to ppb	
		0°C	20°C	0°C	20°C
NO_2	46	2.05	1.91	0.49	0.52
NO	30	1.34	1.25	0.75	0.80
HNO_3	63	2.81	2.62	0.36	0.38
O_3	48	2.14	2.00	0.47	0.50
SO_2	64	2.86	2.66	0.35	0.38
CO†	28	1.25	1.16	0.80	0.86

ie, to convert ppb of SO_2 at 0°C to µg/m³ multiply by 2.86

† for CO the factors apply to the more commonly used conversions of ppm and mg/m³

Appendix 6

Membership of the Committee on the Medical Effects of Air Pollutants

Chairman Professor S T Holgate, BSc, MD, DSc, FRCP, FRCPE

Members Professor H R Anderson, MD, MSc, FFPHM
Professor J G Ayres, BSc, MD, FRCP
Professor P G J Burney, MA, MD, MRCP, FFPHM
Dr M L Burr, MD, FFPHM
Professor R L Carter, MA, DM, DSc, FRCPath, FFPM
Professor B Corrin, MD, FRCPath
Professor A Dayan, MD, FRCP, FRCPath, FFPM, FFOM, FIBiol
Dr A Gibbs TD, MB, ChB, FRCPath
Professor R K Griffiths, BSc, MB, ChB, FFCM
Professor R M Harrison, PhD, DSc, CChem, FRSC, FRMetS, FRSH
Mr J F Hurley, MA
Dr D Purser, BSc, PhD
Dr R J Richards, BSc, PhD, DSc
Professor A Seaton, MD, FRCP, FFOM
Professor A Tattersfield, MD, FRCP
Mr R Waller, BSc
Dr S Walters, BSc, MRCP, MFPHM

Secretariat of The Advisory Group on the Medical Aspects of Air Pollution Episodes (MAAPE) and the Committee on the Medical Effects of Air Pollutants (COMEAP)

Dr R L Maynard, MRCP, MRCPath, FFOM, FIBiol, (MAAPE and COMEAP)
Mrs K Cameron, BSc, MSc, (MAAPE: 1990-1993)
Dr J Greig, MA, DPhil, (MAAPE: 1993-1996)
Dr A Wadge, BSc, PhD (COMEAP: 1992-1994)
Mrs A McDonald, BSc, MSc (COMEAP: 1994-1996)
Dr H Walton, BSc, DPhil (COMEAP: 1996-)
Miss J P Cumberlidge, BSc, MSc (MAAPE and COMEAP)
Mr J D Raghunath (COMEAP 1992-1996)
Mr J Crook, BA (Econ), MA (COMEAP 1996-)

Appendix 7

Bibliography

A short list of major official and other sources is provided. These sources will allow the interested reader to pursue the subjects discussed in this Handbook in greater depth and will provide an introduction to the original literature. All the reports illustrated in this Handbook are included below.

1.	Department of Health. Advisory Group on the Medical Aspects of Air Pollution Episodes. First Report. Ozone. London: HMSO, 1991.

2.	Department of Health. Advisory Group on the Medical Aspects of Air Pollution Episodes. Second Report. Sulphur Dioxide, Acid Aerosols and Particulates. London: HMSO, 1992.

3.	Department of Health. Advisory Group on the Medical Aspects of Air Pollution Episodes. Third Report. Oxides of Nitrogen. London: HMSO, 1993.

4.	Department of Health. Advisory Group on the Medical Aspects of Air Pollution Episodes. Fourth Report. Health Effects of Exposures to Mixtures of Air Pollutants. London: HMSO, 1995.

5.	Department of Health. Committee on the Medical Effects of Air Pollutants. Non-Biological Particles and Health. London: HMSO, 1995.

6.	Department of Health. Committee on the Medical Effects of Air Pollutants. Asthma and Outdoor Air Pollution. London: HMSO, 1995.

7.	Department of the Environment. Expert Panel on Air Quality Standards. Ozone. London: HMSO, 1994.

8.	Department of the Environment. Expert Panel on Air Quality Standards. 1,3-Butadiene. London: HMSO, 1994.

9.	Department of the Environment. Expert Panel on Air Quality Standards. Benzene. London: HMSO, 1994.

10.	Department of the Environment. Expert Panel on Air Quality Standards. Carbon monoxide. London: HMSO, 1994.

11.	Department of the Environment. Expert Panel on Air Quality Standards. Sulphur dioxide. London: HMSO, 1995.

12.	Department of the Environment. Expert Panel on Air Quality Standards. Particles. London: HMSO, 1995.

13.	Department of the Environment. Expert Panel on Air Quality Standards. Nitrogen dioxide. London: HMSO, 1996.

14.	Department of the Environment. Quality of Urban Air Review Group. Urban Air Quality in the United Kingdom. First Report. London: Department of the Environment, 1993.

15.	Department of the Environment. Quality of Urban Air Review Group. Diesel Vehicle Emissions and Urban Air Quality. Second Report. London: Department of the Environment, 1993.

16.   Department of the Environment. Quality of Urban Air Review Group. Airborne Particulate Matter in the United Kingdom. Third Report. London: Department of the Environment, 1996.

17.   Department of the Environment. Photochemical Oxidants Review Group. Ozone in the United Kingdom. Third Report. London: Department of the Environment, 1993.

18.   World Health Organization. Air Quality Guidelines for Europe. WHO Regional Publications, European Series No 21. Copenhagen: World Health Organization, 1987.

19.   Anderson HR, Limb ES, Bland JM, Ponce de Leon A, Strachan DP, Bower JS. The health effects of an air pollution episode in London, December 1991. London: St George's Hospital Medical School, 1994.

20.   Medical Research Council. Institute for Environment and Health. IEH Report on "Air Pollution and Health: Understanding the Uncertainties". Leicester: Institute for Environment and Health, 1994. *IEH (1994) IEH Report on Air Pollution and Health: Understanding the Uncertainties* (Report R1). Leicester, Instritute for Environment and Health.

21.   Medical Research Council. Institute for Environment and Health. IEH Report on "Air Pollution and Respiratory Disease: UK Research Priorities". Leicester: Institute for Environment and Health, 1994. IEH (1994) *IEH Report on Air Pollution and Respiratory Disease: UK Research Priorities* (Report R2). Leicester, Institute for Environment and Health.

22.   IEH (1996) *IEH Assessment on Indoor Air Quality in the Home: Nitrogen dioxide, Formaldehyde, Volatile Organic Compounds, House Dust Mites, Fungi and Bacteria* (Assessment A2). Leicester, Institute for Environment and Health.

23.   Medical Research Council. Institute for Environment and Health. IEH Assessment on Indoor Air Quality in the Home. Leicester: Institute for Environment and Health, 1996.

24.   Berry RW, Brown VM, Coward SKD, Crump DR, Gavin M, Grimes CP, Higham DF, Hull AV, Hunter CA, Jeffery IG, Lea RG, Llewellyn JW, Raw GJ. Indoor Air Quality in Homes: Part 1. The Building Research Establishment Indoor Environment Study. Watford: Building Research Establishment, 1996.

25.   Berry RW, Brown VM, Coward SKD, Crump DR, Gavin M, Grimes CP, Higham DF, Hull AV, Hunter CA, Jeffery IG, Lea RG, Llewellyn JW, Raw GJ. Indoor Air Quality in Homes: Part 2. The Building Research Establishment Indoor Environment Study. Watford: Building Research Establishment, 1996.